Computer Science Why

Computer Science Why is a book that presents a straightforward, curiosity-based approach to filling in the blanks around common, introductory material taught in Computer Science (CS) classes. If you have ever wondered "why is that?" about a programming phenomenon, historical tidbit, or common terminology, this book may have an answer that explains it.

Questions include:

- Why do computers have trouble with floating point numbers?
- Why is it called Bluetooth? Why does it use that symbol?
- Why is it called bootstrapping?
- Why don't modifications to function parameters persist once the function returns?
- Why do we use the QWERTY keyboard?
- ... and many more.

Perfect for both students and IT professionals, this book provides the clear "why" to answer so many unstated CS questions. Take any example in this book, from base 2 to time zones, and you will have an answer that cements your insight on the "why" of the topic, reinforcing your rote memorization with deeper understanding and insight.

Rachael Little has a lifelong passion for computer technology and is currently finishing a PhD in Computer Science at the University of New Hampshire.

Christopher Little has a lifelong career in (and passion for) software and computer technology. He has been a developer and/or managerial at organizations such as NASA's Johnson Space Center, IBM, Health Language, BMC Software, and Gartner Group. He has started (and sold) several companies.

Len Little has a dedicated interest in linguistics and computer security, and is currently getting an undergraduate degree in Computer Science and Cybersecurity at the University of Liverpool.

Computer Science Why
Answers to Computer Science Questions They Don't Cover in Class

Rachael Little, Christopher Little, and Len Little

CRC Press
Taylor & Francis Group
Boca Raton London New York

CRC Press is an imprint of the
Taylor & Francis Group, an **informa** business
A CHAPMAN & HALL BOOK

Designed cover image: Rachael Little, Christopher Little, and Len Little

First edition published 2026
by CRC Press
2385 NW Executive Center Drive, Suite 320, Boca Raton FL 33431

and by CRC Press
4 Park Square, Milton Park, Abingdon, Oxon, OX14 4RN

CRC Press is an imprint of Taylor & Francis Group, LLC

© 2026 Rachael Little, Christopher Little, and Len Little

Library of Congress Cataloging-in-Publication Data
Names: Little, Rachael author | Little, Christopher (Software developer)
author | Little, Len (Computer science student) author
Title: Computer science why : answers to computer science questions they
don't cover in class / Rachael Little, Christopher Little, and Len Little.
Description: First edition. | Boca Raton FL : CRC Press, 2026. |
Includes bibliographical references.
Identifiers: LCCN 2025044645 (print) | LCCN 2025044646 (ebook) |
ISBN 9781032856759 hardback | ISBN 9781032856742 paperback |
ISBN 9781003519379 ebook
Subjects: LCSH: Computer science
Classification: LCC QA76 .L535 2026 (print) | LCC QA76 (ebook)
LC record available at https://lccn.loc.gov/2025044645
LC ebook record available at https://lccn.loc.gov/2025044646

ISBN: 978-1-032-85675-9 (hbk)
ISBN: 978-1-032-85674-2 (pbk)
ISBN: 978-1-003-51937-9 (ebk)

DOI: 10.1201/9781003519379

Typeset in Kepler
by codeMantra

Contents

Introduction

Undergraduate students learning various topics in Computer Science (CS) will encounter all sorts of conventional, prescriptive knowledge which they often must learn and memorize by rote force. Oftentimes, this knowledge is structured upon low-level, architecturally deep, and historical foundations that are impossible to fully convey and explain in introductory classes. Students must make do in learning numerous contexts without any additional context for the term or concept (the "why" of it).

This book provides a clearer "why" behind many CS topics, so that instead of simply memorizing terms, the reader will be able to understand and utilize their context. Take any example in this book, from Bluetooth to floating point math to time zones, and you will read an answer that cements your insight on the "why" of the topic. The idea is to replace memorization with insight.

Our charter in writing this book was simple: encourage students to expand their understanding of CS both by presenting the questions many are hesitant to ask in class, and then by shedding light on CS internals with the answers. We provide an answer that provides a revealing "ah-ha!" moment—specific, digestible explanations that an undergraduate CS reader can not just remember, but also use. Once you read the answers we have provided here, the reasons behind the topics are, we believe, plain yet vibrant and unforgettable.

This book is not a replacement for CS course, work, or study. It supplements them. It is intended to augment those textbooks and other reference books and manuals. It is, however, freestanding as a source of answers and insights. It can be used on its own, but we intentionally made the answers only one layer deep. Thus, each answer is only the answer to the specific question, although some answers (and questions) raise other questions (and answers).

Writing the answers in this book has been a fascinating process. It has often resulted in many draft pages (way too many!) to an answer as we'd dig into the topic. Those draft pages would then need to be boiled back to creating the specific answer. Even well-known answers often became more complicated at first when explored. Trivial example: the original QWERTY entry grew to multiple pages about the early typewriter wars and their feuding keyboard designs (some were quite remarkable!). These were then all removed as being essentially irrelevant to addressing "why QWERTY," and we focused on the winner we have today. That is, the keyboard layout (a Latin-based alphabet) that most everyone uses today.

We often found fascinating insight and detail which were irrelevant to the specific answer and, in turn, to the intent behind this book. While these were fascinating asides, they had to be pruned and edited out over time.

Our operating goal can be summed up with this question:

What makes this answer most useful and memorable to a new CS student?

Or, short form:

What's the "ah-ha!" answer for this question?

We recognize that nearly any answer we provide will be debated or perhaps considered wrong by someone, somewhere. We know that some will not agree with how we answered some questions.

Every answer has been extensively researched, reasoned, and reviewed. Sometimes a source had to be discounted in favor of others when there was a preponderance of agreement (or lack of it). Some had claims which could not be verified and then had to be removed. Sometimes we decided to follow the consensus view. Sometimes a topic is just a mess of debates, and we settle on favoring the logical answer. Where relevant, these reasons are indicated in the answers. We have not seen these answers in print before, nor have we seen them answered in the spirit we have pursued the question (providing the "ah-ha!").

Most of the answers have the following general structure: a first paragraph which generally provides an immediate answer to the question. The rest of the answer is usually a follow-on section which provides more nuance and detail around the answer (sometimes historical, sometimes explaining hidden complexity or detail).

We invite feedback. We look forward to it, in fact. We expect there will be lively discussion. Some of these answer topics have long-standing debates around them, and we've found that the debate perspectives are usually instructive.

Please email us with feedback, comments, and questions about the book at: **cswhy@gmail.com**

This book also has a companion website at www.cswhy.com. More in-depth details will be provided about answers in this book, how we arrived at certain decisions, the wording of certain answers, and more. The entire reference and bibliography will also be live-linked there, so if you found links in the book and want to delve deeper, that'll be the place to go.

To our knowledge, no book like this exists. The idea for this book started from a casual conversation between the three authors about CS education. Len was getting an undergraduate degree in CS, Rachael was finishing up a PhD in CS, and Christopher has spent a career in computers and software.

We were sitting in the family living room. Len was back home on holiday break, and we were discussing the state of CS education and why some of it seems unnecessarily hard. It began with what Len was experiencing in undergraduate CS studies and its frequent requirement of rote memorization. Then, Rachael's experiences of teaching assistantship work in pursuit of her PhD studies where students were often required to just memorize things. Rachael found that if she explained certain things as to why they worked the way they did, the students understood and then performed much better.

In most cases, these CS questions—that students were memorizing answers for—had perfectly good and compelling explanations or "why"s. Explanations make the topics clear, compelling, and memorable, without the need for rote memorization. When you understand why a thing is the way it is, you know something foundational about it. Everything builds from there.

We wondered how many such questions there were and started a list. It soon exceeded 20 questions and, by the end of a month, there were well over a hundred. As of this writing, there are over 150 questions in our list. So then we set about writing some example answers to these mostly why-based questions (why are floating point numbers hard to work with, why is it called Bluetooth, etc.). We started with four answers.

When we looked over the answers and realized we had another 100 in the wings, we realized it was a book.

Acknowledgments

The authors would like to thank our publisher, Randi Slack, for all her help and encouragement in the creation of this book.

Len thanks Claire Wong for her unending support during this process (as well as over the last 10 years) and Sashitra Trivinnaka Murli for her proofreading, including the pushes to finally finish answers after much procrastination!

Rachael would like to thank numerous professors and mentors, who expanded her inner world view of Computer Science understanding and encouraged her to press on in her studies, in no particular order: Dr. Dongpeng Xu, Dr. Huw Read, Professor Arvind Narayan, Dr. Elizabeth Varki, Dr. Jeremy Hansen, Professor Kris Rowley, Dr. Matthew Bovee, Dr. Henry Collier, and Dr. Addie Armstrong.

WHY ARE TOPICS LIKE MATH, STATISTICS, AND FORMAL LOGIC FEATURED SO HEAVILY IN COMPUTER SCIENCE EDUCATION?

Given that Computer Science (CS) and related degrees are actively evolving in schools both across the United States (where the authors are based) and globally, the topics in this entry may or may not be very relevant to you at the time of reading. Today, CS education is split into classes of specialties like Computer Security, Digital Forensics, IT Infrastructure, and Software Development, which may or may not require all the math and theory-based courses listed below. Regardless, it's worth describing how these topics have been and continue to be essential to the wide ranging fields of CS.

Below, we list them by area, and describe their early influences on computer development and how they tie into the field today.

Calculus

You've likely had to take this class at your school's math department, and may or may not have had opportunities to tie it into your CS-related material directly. Calculus provides essential tools for numerous areas of software development.

Calculus is essential for various simulations and computer graphics. Describing the rate of change and momentum, which requires derivatives, is essential for things like rendering curves and animating motion. When you play a video game on your computer, and you see an object bounce, change momentum, fly, or fall and hit the ground with a realistic appearance of motion, calculus is used in the code that animates it.

The calculus of Fourier transforms (or waveform analysis) allows us to apply computers to a wide variety of problems requiring time-frequency analysis. This is significant math for the world of signal analysis, where applications include examples such as tuning musical instruments, understanding sound composition, measuring the distance to a star, or even determining the color composition in a beam of light. Basically, Fourier transforms encompass the analysis of any composite event, whether the material to be analyzed is visual, aural, physical, chemical, or thermal.

Calculus is an important concept for artificial intelligence (AI) and machine learning (ML), as it's used in optimization functions, which are calculations that gradually inform and improve the AI or ML model's performance at a particular task. Calculus provides the theory to adjust the performance of various functions in relation to certain data, which is fundamental to neural networks' and other learning models' ability to adjust their accuracy by modifying unique values called "weights." Importantly, calculus also allows for chaining these calculations together, which increases the depth and complexity of the weights that AI and ML models can use (and in turn, the depth and complexity of their tasks).

Although at times a dreaded class for CS students, calculus concepts are absolutely essential for many areas of our digital world.

Linear Algebra

Linear algebra comes in tremendously handy in any field that requires matrix operations, and as it turns out, there are many of them! Modern CPUs actually have dedicated instructions for performing arithmetic operations on matrices quickly, as they are such commonly required structures in varying areas of software development.

Computer graphics is, once more, a field where math like this comes into a great degree. Consider your typical video game where you're in a 3D world, moving around; every time you or another object moves around the screen, each pixel of that object must be re-rendered. Information about the location for that object is stored as coordinates, and when you move around in a video game, or rotate the camera, the computer must re-render everything on the screen using its coordinate info relative to the camera's. As it turns out, visual transformations can be applied quite easily to coordinate info stored in matrix form. Rotations, horizontal and vertical movement, stretching and compression—almost any kind of force acting on an object in a rendered area can be expressed as an operation upon a matrix.

ML and AI make extensive use of this as well, by storing and manipulating data in vector- and matrix-based formats. Matrixes are an efficient way to store multidimensional information about model weights and parameters, and linear algebra is the foundation on which these methods work.

Discrete Math

Discrete math includes the study of Boolean logic, reasoning, and set and graph theories.

One of the most well-studied domains of discrete math may be Boolean logic (sometimes referred to as Boolean Algebra). Boolean logic is derived from George Boole's 1854 book on Logic (where its very long title is usually abbreviated to "Laws of Thought").

The value of Boole's work is hard to overstate for CS, although at the time it was published, it was not seen as generally useful (particularly given there were no computers!). Boole's work was a breakthrough notational codification of symbolic or logical reasoning. Unlike Venn's or Dodgson's logic diagramming innovations, it allowed for direct encoding of propositions to be used in a logic notation form. It contains a small set of operator rules and operators for transcribing logical expressions, allowing even for the representation of traditional logical forms such as Modus Ponens or Modus Tollens.

Basically, what we know today as the AND, OR, and NOT gating computer circuits were encoded functions in Boolean logic (and, later, NAND, NOR, etc.). For the second half of the 18th century, Boole's logic system became widely adopted and continues to this day to be a means of representing logical propositions and resolving truth conditions within the study of philosophy, but until computers, it was not generally otherwise well known or used.

A key insight is that Boolean functions of math and identities build on Boolean Truth Tables (a matrix of states showing truth conditions, true or false, listed with a final true or false outcome being the result of the operation).

In the 1930s, a dual engineering and mathematics graduate named Claude Shannon noticed that these truth tables allowed Boolean logic to be applied to computational systems, where circuit states were generally implemented as "on" or "off"; i.e., in binary. More powerfully, for information theory (which Shannon is generally credited with creating) and computation theory (see below), Boolean logic also gave the ability to manipulate and derive truth states for symbolic values (see below).

Operators AND, OR, and NOT Truth Tables using variables A and B

AND operation

A	B	Result
False	False	False
False	True	False
True	False	False
True	True	True

OR operation

A	B	Result
False	False	False
False	True	True
True	False	True
True	True	True

NOT operation

A	Result
False	True
True	False

It is pretty easy to see Boolean logic as being directly related to CS at the circuit level—after all, the electronic circuits that make up your PC or laptop hardware are based on that very logic! When designing circuits, engineers use truth tables to express functionality at the bit level (1s and 0s), and this is based on using Boolean logic. A small handful of functions (implemented as types of gates to control electrical flow) allow for massive complexity to be delivered at human scale as the computer performs in some cases millions of these computations in a second.

Boolean logic is also essential for the mathematical and symbolic underpinnings of programming languages at all levels, from machine code to assembly to higher-level generation-defined languages, which we cover a bit more in the next section.

Theory of Computation

This usually spans a range of topics in the actual "Science" part of "Computer Science," covering things like finite automata, computability theory, and formal languages. Here you'll likely face exercises in formal logic (including using the above-mentioned Boolean logic), proofs, evaluating regular expressions, and determining the valid arrangements of a language given a grammar specification. Although often deeply abstract in a classroom setting, to the point where professors have published papers on devising effective teaching examples for this class, these topics make up the underpinnings of all functioning programming languages and software we use today.

Among other items in computation theory, finite automata, which are used to describe the state of a system given a series of possible events, can model and allow testing for numerous real-world systems. This includes, for example, oddly subtle things like the optimal sequencing and cross-coordination of blocks of traffic signal lights.

Formal language theory first appears in compilers, where the use of parsers and lexical analyzers ("lexers") transforms user input into the specific code and instructions in a language specification. It plays a bedrock role in such processes. It can then be supplemented with logic for the development and testing of high-level programming language features, such as anonymous functions and lambdas.

Computability theory, which has to do with determining the complexity and feasibility of computing various models, has modern-day applications ranging from program verification and testing to determining the complexity of an algorithm before implementing it (to evaluate in a setting where the best time and memory performance are of the utmost importance). The proofs around these models rely on and extend the Boolean logic we mention from the previous section, building logical statements and outcomes that can be formally verified.

So, from using simple truth tables and Boolean logic, with outcomes of 1 (True) or 0 (False), you may encapsulate upward with language theory into a variety of higher symbolic representations, such as an example of the logic classic (i.e., modus ponens).

> All men are mortal (TRUE)
> Socrates is a man (TRUE)
> therefore, Socrates is mortal (TRUE)

This kind of theoretical foundation can be applied to software and computational statements, then giving programmers the ability to create flows and constructs that can, at scale, emulate human systems and processes, from accounting to rocketry.

Number Theory

This area is massively important when understanding modern cryptography. Number theory classes cover, among other things, the basic concepts that make up public key encryption (or PKI), modular arithmetic, one-way functions, and digital signatures. When asked to come up with a scheme to encode a message, most people might talk about those which require a passkey, a special secret knowledge, or another agreed-upon value that has to be shared with the message recipient. But modern encryption uses the topics described in number theory to be completely passkey-less between sender and receiver, which is a must for secure communications today, where computers must initiate new connections between each other without being able to exchange a password or any secret info beforehand.

Another entry in this book (see the question "Why is public key private key encryption so popular in online communications?") explains this as well, but for a little detail here: by applying number theory's concepts around prime numbers and modular arithmetic, it is possible to generate two keys, one private and one public, where the public key can be used to encode a message, and only the private key (kept secret from everyone except the recipient of the message) can decode it. This encryption is strong enough that it would take years to attempt to brute force and figure out the private key for decryption. This type of encryption is fundamental to the security of everything we do over the Internet as it keeps potential attackers from being able to monitor our communications (sending our login password to a banking website, exchanging private messages, and more) and being able to pick up information just by observing the data being sent back and forth across the network.

Number theory also provides the toolset for things like hashing and digital signatures, which allow for the verification of digitally stored information. Considering that digital data is, essentially, 1s and 0s when you look under the hood, and the potential challenges prompted by transmitting over the Internet, verifying that your data is correct and from the source you expect is actually a vitally important and nontrivial task.

Statistics

One area you might not expect statistics to come in handy is also in ML! When training and learning on data, networks and other ML models must continually make educated guesses about what adjustments to try next. Statistical models such as linear regression, hypothesis testing, and analyzing the distribution of data are all tools essential to ML.

Data science, as applied statistics, is one application that may be obvious, but is still very worth mentioning here. Collecting and visualizing information in meaningful ways is a skill just about every business and scientist requires, whether it be for tracking customer shopping habits, identifying trends in product performance, or just communicating information about trends quickly and effectively to others who might not be as well versed in statistics.

FURTHER READINGS

Bunde, D.P. (2019). Using real examples to motivate automata theory. *J Comput Sci Coll,* *35*(5), 28–36.

freeCodeCamp: https://www.freecodecamp.org/news/boolean-algebra/

George Boole: https://georgeboole.com/boole/life/biog/

IDSIA: https://people.idsia.ch/~juergen/leibniz-father-computer-science-375.html

Knox: https://faculty.knox.edu/dbunde/pubs/practical-automata.pdf

MathOverflow: https://mathoverflow.net/questions/10334/what-practical-applications-does-set-theory-have

Medium: https://medium.com/effortless-programming/the-missing-introduction-to-calculus-for-ai-3d5e8df6efa3

Project Gutenberg: https://www.gutenberg.org/files/15114/15114-pdf.pdf

ScientificAmerican: https://www.scientificamerican.com/article/claude-e-shannon-founder/

University of Cambridge: https://www.cl.cam.ac.uk/teaching/1213/DiscMathII/DiscMathII.pdf

Architecture

Computer architecture generally has to do with the underlying representation of program memory, data, and related structures. We cover questions about program behavior and programming quirks that have answers rooted in this low-level area.

DOI: 10.1201/9781003519379-1

WHAT DOES 64 BIT REFER TO?

The term "bit size" refers to the maximum memory size that a computer can support. 32-bit computers, for example, can only use 4 GB of RAM. 64-bit computers can use a substantially larger amount.

A 64-bit computer has CPU registers sized 64 bits long, or 8 GB. This means that the maximum number any given register can hold is equal to 2^{64}, or in decimal, 18,446,744,073,709,551,615. So, how does this affect computer memory?

Everything in memory exists at a fixed address. In order to access data, programs must be able to refer to them by address. The maximum value, and thus address, which can be formed in active computer memory to do so is determined by register size—and this is where 64 bit comes in. On 64-bit machines, registers are sized at 64 bits long, which determines the maximum address they can form. This is why 32-bit computers can only utilize such a limited amount of RAM—their CPU registers physically cannot form addresses large enough to access memory beyond 4 GB. (See the question "Why do base 2 number systems feature so prominently in Computer Science?").

In case you're wondering: why do we refer to them in bits? This has heavily to do with the binary numbering system. For more info on that, please check out our other answer, "Why do base 2 number systems feature so prominently in Computer Science?"

WHY DO BASE 2 NUMBER SYSTEMS FEATURE SO PROMINENTLY IN COMPUTER SCIENCE?

The binary numbering system, or base 2, is used quite extensively throughout computer architecture. At times, octal (base 8) or hexadecimal (base 16, which is easily recognizable by its use of letters) are also used, but you may notice these bases are also powers of 2—there's a good reason for this!

For a brief refresher, the "base" of a numbering system refers to how many digits it uses. Decimal, or base 10, is the numbering system you are likely most familiar with, and the system used in schools and workplaces around the world. It uses 10 digits—"0" through "9." Base 2 only has "0" and "1." The base of a number ultimately determines its notation of a particular value—"11" in binary is equivalent to 3 in decimal.

On the computing side, hardware must store and perform computations using electrical signals, and binary corresponds to this extremely well. "0" and "1" can be easily transmitted as "on" or "off" in an electrical circuit, via voltage changes over wires, or when representing the bit status on CPU registers. Although modern programming projects and applications typically have plenty of built-in support for conversion to the more human-readable decimal system, it's still helpful to have an awareness of binary and hexadecimal, as these are the closest, most accurate representations of the actual bit values stored in CPU registers, on the stack, or on the heap.

Earlier, we mentioned that hexadecimal and octal are powers of 2—this is a very handy property! This means that binary numbers can easily and compactly be translated to these forms, 4 and 3 bits at a time, respectively.

Let's look at a few examples. 0111 in binary is equivalent to 0x7 in hexadecimal. 10100111 (1010+0111) is equivalent to 0xA7 in hex. Every 4 digits in binary can be compactly encoded into a single hex digit, no matter how large or small the number grows. Conversion from binary to decimal and vice versa is not so simple—and this is why hex and binary are often used interchangeably for displaying base 2 numbers. Hex is a much more compact and directly proportional way of storing binary, and is used quite popularly as the format of data representation for computing systems.

Octal is less commonly used in everyday programming and debugging as it is not quite as compact, but early on in computer development, it was the primary numbering system for machines that featured smaller bit systems, such as the UNIVAC 1050, which had a maximum instruction size of 16 bits, or the PDP-8, which was a 12-bit machine. Although octal is rarely used in modern systems now, its presence still has a holdover on many popular command line tools. Octal options are available for output within numerous debugging and binary display tools, and the Unix/Linux permission settings make use of octal, as the possible settings happen to fit well into octal and there has never been

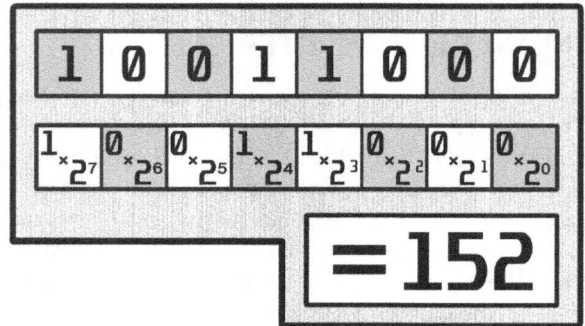

a need to change it. Because any system that can display hex can also display octal, there is no pressure to remove it from popular modern tools.

Now if you're wondering: what does 64 bit mean when it comes to processor versions, and what's the difference between a 32-bit and 64-bit computer? Turn to the answer for the question titled "What does 64 bit refer to?" to find out!

WHY DOES BIOS/UEFI EXIST AND BEHAVE THE WAY IT DOES?

The Basic Input/Output System (BIOS)—later generation is called the Unified Extensible Firmware Interface or UEFI—is firmware to initialize the microcomputer hardware and assist in loading of the operating system. The BIOS phase boots up the underlying hardware on the computer, which then loads the operating system, or, in the case where multiple operating systems are installed (dual boot or multiple hard drives), allows the user to select which OS they'd like to boot into. It comes preinstalled on motherboards and microcomputer hardware to perform hardware initialization during the booting process.

In addition to acting as a pre-boot phase where the OS can be selected, it is used during POST to establish that the basic hardware (keyboard, disk drive, monitor, HDD) is initialized and operational for communication as the computer is tested through the POST sequence. These things must be fully operational before the OS can be loaded (otherwise, the user ends up with a system they can't control, or even fix after booting into the OS!), which is why these checks exist separately before loading the OS.

Exclusively microprocessor-based and coined by Gary Kildall in the late 1970s, BIOS originally referred to the Basic Input/Output System in the CP/M OS. Its usage was adopted by IBM in 1981 and has since become widespread. The UEFI is an open standard created to replace BIOS but do the same functions, while providing for faster boot times, larger images, file support, and a user interface.

WHY IS CLOCK SPEED IMPORTANT IN PROGRAMMING?

Clock speed generally refers to the number of cycles per second in which the motherboard microprocessor (the "central processing unit" or CPU) is designed to execute instructions, with the number most often being expressed in terms of a frequency or counter.

It is important to note that while clock speed is itself measured in elapsed time, it is not in fact a measurement of elapsed time but of computer CPU executions (which occur over an elapsed time). Thus, "clock speed" is used as a metaphor for the performance of the CPU and not the time that might be seen on, say, a wall clock.

Although computers may execute instructions very quickly, they execute those instructions only at the intervals provided, exactly at the tick of the system's CPU. Since a step only happens at the tick, clock speed is thus the key determinant for the performance of the overall computer. It determines *when* the processor will execute something, while the programming instruction determines *what* the CPU will execute.

In other words, clock speed determines how quickly a computer computes, the optimal sorts of signals or events the computer can sample or monitor, how it communicates with internal components such as storage, what the computer can output, and the frequency with which it can communicate with other devices.

Thus, you find gamers will often "overclock" a CPU—artificially increase CPU execution frequency above manufacturer's specifications—to improve the execution of the online game on their machine and thus improve the responsiveness and performance with which they can play such games. Overclocking can become a competitive gaming advantage in that scenario. The downside to overclocking is that it has the potential to overheat the CPU due to running it at beyond-specification frequencies. Overheating the CPU can be damaging to the CPU.

The timing of a CPU also plays a direct role in what the overall device is able to monitor and process. The sampling and monitoring must align with CPU executions, which may or may not overlap well with the source being sampled or monitored.

Sampling Trivia: In older movies that show a traditional Cathode Ray Tube (CRT) display (not LED, plasma, etc.) and were recorded on film (not digital), you can often see the sweep of the monitor display because of the lack of synchronization between the film—usually recording at a frame rate of 24 frames per second—and the CRT screen sweep. A non-digital example of this effect (called "the wagon-wheel effect") is often seen in westerns when wagon wheels roll forward, but when the rotations reach and then exceed the capture frame rate of the film camera, the spoked wheels will appear to stop instantly, then start rotating in reverse due to the camera sampling at a now lagging rate.

Measuring an event is usually a function of how *often* you can sample it. You cannot accurately measure what you cannot accurately sample. If what you are monitoring or sampling occurs at higher rates than your CPU, you will miss any data points that occurred while the CPU was not able to process them. This can have serious consequences.

Here is an example: For a typical multipoint 'QRS' heartbeat waveform, you need to capture the whole waveform to understand the nature of the patient's heartbeat (and thus state of the patient's heart). If the device cannot process that in time with the events, then one or more of the points will get missed and the result, for that waveform, is virtually useless for clinical evaluation purposes.

The effects of clock speed are not to be confused with latency, although the effects from observing a device with a higher execution speed from a device with a lower execution speed can appear to behave like latency because of the experience of missed signals or missed sampling of events.

FURTHER READINGS

Early Apple Steve Jobs talking about computing and in turn (without stating it this way)
    the effect of clock speeds: https://stevejobsarchive.com/exhibits/objects-of-our-life
How to learn your clock speed: https://www.wikihow.com/Check-CPU-Speed
https://www.intel.com/content/www/us/en/gaming/resources/cpu-clock-speed.html

HOW DO DEBUGGERS "PAUSE" PROGRAMS? HOW DO DEBUGGERS CATCH THINGS LIKE SEG FAULTS?

The short answer is: they don't really pause programs, at least not directly! Debuggers perform their duties with a tremendous amount of help from the operating system and CPU. What looks like pausing, single-stepping, returning until execution, and stopping on fault or on exceptions in real time is actually a very intricate dance between the debugger, the OS, and a common understanding of OS API calls and careful state management. This answer assumes you already have a basic familiarity with what debuggers are and how they work; if not, this is a good time to do a quick search on some basic tutorials.

Before digging into how debuggers manipulate programs, let's recap on how programs are typically executed. Programs are stored on your machine in a binary format, or in other words, as a set of assembly instructions called opcodes. When the program is started, it is loaded into active system memory—meaning the instructions, data, and other important information about the program are copied into your RAM for faster execution. When the CPU executes these instructions, it does so in a series of steps called the fetch–decode–execute cycle, in which the CPU will fetch the next instruction from the program, decode it (or, in other words, calculate any addressing info needed), and then execute it. It does this in a steady and continual sequence.

So, when we consider how to pause a program for debugging, we need some way to tell the CPU to pause execution, and we need to do it in such a way that it doesn't change the program behavior, mess up its active memory, or interfere with the CPU's overall operations. After that, once we're done reading all the information we want from this paused state, we need a way to signal to the CPU to continue. In addition to all of this, we need a way for the debugger to gain access to the target process' memory—as typically, process memory is private. This is a complex problem to solve, and early on in computer architecture development, computer architects realized it would be helpful to build in support for pausing programs within the CPU instruction set itself. As a result, popular operating systems today, including Windows, Linux, and MacOS, all have a set of API calls which debuggers can use to interact with programs. Although these API calls vary slightly between operating systems, they all cover the same basic actions.

Here's how it works:

Step 1: Gaining Access

First, the debugger must gain permission to manipulate a program, as typically, process memory is private and inaccessible by other processes. It must issue a request to do so with an OS API call, where it asks for permission to become the designated debugger of a process. The OS returns information about the process, like its current address space and name. The debugger can use the OS API calls in two ways: either to attach to an already running process or to start the process from scratch. This also registers the debugger with the OS so that it knows what program is acting as the current debugger—the importance of which will become clear in the next step.

Step 2: Setting Breakpoints

When you set a breakpoint within a debugged process, you are essentially marking the program at a given line and asking for it to stop executing at that location. When you start it, the OS loads the program into memory and the CPU executes all instructions (in the form of bytecode), in order. We mentioned earlier that the CPU has a fetch–execute cycle in which it runs through each instruction of a given program—and somehow, therefore, it must be signaled to stop at a selected point. In modern systems, there is a dedicated assembly instruction which indicates to the CPU that there is a breakpoint; for the x86-based CPUs, this is called "INT3." When you set a breakpoint at a given location, the debugger actually writes the opcode for this into the program's active memory. When the CPU encounters one of these instructions, it immediately halts, and then passes state information and control back to the currently registered debugger (which it is aware of because we registered it using that API call to gain ownership of the process!) At this point, the debugger will wait for user input, for instance, you hitting the "run" button again.

If by now you're thinking, "I've looked at the assembly view in my debugger. Why do I not see the INT3 instruction there when I place a breakpoint?" Debuggers hide this in their output when showing assembly instructions so as not to be confusing to users. It's typically unnecessary to include the INT3 instruction embedded in assembly for everyday debugging purposes, as it doesn't affect program execution, and the average user isn't aware that it needs to be there for debugging.

Note that what we've described above applies to software breakpoints. Hardware breakpoints perform the same purpose, but are implemented by the use of special registers on the CPU dedicated to storing breakpoint locations. These are a bit faster than software breakpoints, as they don't require modifying instructions in memory. They also have an advantage (or disadvantage, depending on what your goal is) in that by not modifying the process' memory space, they can't be as easily detected by programs that have protections against being analyzed (including gaming and commercial software, as well as malware). The downside is that there are only so many hardware breakpoints that can be set at a given time, whereas software breakpoints can be nearly unlimited in number. Many debuggers support using both kinds at once.

SO HOW DO DEBUGGERS CATCH THINGS LIKE FAULTS, PROGRAM CRASHES, AND EXCEPTIONS?

This topic is a bit complicated! But the short of it is that it depends on the type of exception or crash, and what language you're working with. Let's say you have a Python or Java program, and it crashes because you've written over some array bounds. In this case, you'll get a detailed error message and line number, because the backend of those programming languages includes throwing exceptions for particular types of programming mistakes. When you use, say, Java's Array library, and try to place something past the array bounds, Java's backend code will detect that this is invalid, and throw an out-of-bounds exception. In more primitive languages like C or C++ (or in other words, languages that implement very low-level concepts like pointers), when you write past array bounds, you might get a slightly more confusing message, like "stack smashing detected." This is because when you work with the primitive array type in C or C++, you're working with program memory on the stack, and so there are fewer ways to detect memory overwrites. The C/C++ compiler will typically embed code to place special values at select places on the stack (called "canaries"), and periodically check that they haven't been changed. If they have, then this means that an invalid memory write has occurred—like when writing past array bounds.

In both cases, the program output will often show you a stack trace, or history of the entire function call leading up to that error, which it's able to do because of how programs use stack space. If you're familiar at all with the assembly calling convention, recall that every time a function call takes place, the caller address (location where the function was called) is stored on the stack and pointed to by the frame pointer. The result is that wherever you are within the program, however many functions deep, there always exists a chain of pointers leading back up the caller addresses all the way back to main. When a major exception occurs, the program can simply use this information to recreate a map of the function calls that lead to this point, and print it out—hence the name, "stack trace."

That still leaves the more frustrating kind of crash: the segmentation fault, which you'll be familiar with if you've had to do any work with dynamic memory (allocating pointers or objects) in C and C++. With no additional build settings enabled, this will usually just give you a generic error message and very little idea of what caused it. Segmentation faults are a type of memory error where your program writes to an invalid address somewhere outside the program's valid memory range, so that the operating system has to kill the program to keep it from corrupting anything. Because the cause of this crash originates from outside the program, it's often not quite in sync with the instruction that actually caused the problem, and you may not be able to get a stack trace in your crash output. In some cases, because system memory layout can change from run to run, it's even possible to write to an invalid address that happens to be still within program space, and not see any crashes or bugs, and then on the next run, hit an area which causes a segmentation fault. These kinds of faults can be incredibly difficult to debug, as they don't occur consistently, and when they do crash, it might not always be at the same location. Tools like Valgrind, which is specifically built to detect memory problems in C and C++, are helpful for finding these problems before they cause crashes.

Lastly, there's a special class of bug called the "Heisenbug," an extremely finicky phenomenon. This refers to bugs that disappear or appear only when the program is being debugged. This can happen in a number of ways: for instance, someone might notice a problem that occurs during multithreading where two threads reach a deadlock for a particular resource, and then when they pause one or the other via a debugger to examine them, the problem disappears as their timing has been completely changed. Heisenbugs can work the other way around; debugging a complex application which interacts with other processes or network communications could introduce issues in behavior by interrupting the typical flow of operations which don't typically appear.

And so this concludes an overview to the underside of both debugging and exception handling—in short, there's no magical way to pause programs, examine their internals, or glean error information, like with the flick of a wand. These actions are supported by the CPU, the operating system, and a heavy amount of debugging output built in by system architects and language developers alike. Computer Science is magical in many ways, but at the end of the day, there's nothing mysterious under the hood—beneath the surface, effective computing is a huge and complex orchestration of collusion between all the inter-operating parts, and all operating in unison for a common goal.

For more reading on this topic, we recommend checking out these resources:

How Debuggers Work by Johnathan B. Rosenberg: A 1996 textbook on debugging mechanisms and algorithms. Although it is an older publication, it is still highly relevant today.

The Debugging Book (debuggingbook.org)—This is an interactive website which teaches debugging concepts and how to build a debugger in Python.

WHY IS ACCURATE DISASSEMBLY SO DIFFICULT? GIVEN THE DIFFICULTY, HOW DO COMPUTERS "KNOW" HOW TO RUN PROGRAMS CORRECTLY?

Before we fully dive into this answer, a quick recap on the definition of disassembly: the process of taking an executable and accurately extracting its assembly instructions. It's tempting to think that this should be easy, as the series of bytes that make up the executable are the actual assembly language opcodes just one after the other, and therefore disassembling the executable should be as simple as running through each opcode linearly and translating it back. In reality, a number of factors make this quite difficult. Furthermore, if you've ever read much about efforts to create modifications (mods) to games, you might have heard that game production companies insert protections into their software to prevent analysis and disassembly (and thus, the ability to customize games, insert cheats, or steal proprietary code to build and sell similar games). Many other commercial fields use similar protections, but for this entry, we focus on game modifications as an easy-to-portray example. You might also wonder: how do these disassembly tweaks prevent analysis, but not cause problems for your operating system and CPU when they are executing the program?

This will take some context to fully explain, including how programs are laid out, how they are executed, and how disassemblers work. Before we get into any of that, though, we'll start with a few examples of assembly and how even a few short instructions can provide a confusing task for parsing.

Assembly Basics

Assembly (which we'll sometimes call asm for short) is made up of sequences of "opcodes," which each correspond to a CPU instruction like "add," and their operands, the numbers, memory locations, and registers that are used by the instruction. For this entry, we focus on x86 Intel examples, although similar concepts apply to other architectures. These instructions are very precise, and so even simple actions like "add," "subtract," and "jump to a certain location in the program" have many associated opcodes, each one corresponding to variations like "add a register of a certain size to another register," "subtract the value of a certain register from a particular memory location," or "perform a jump at an offset." In Intel x86, for example, the "add" instruction has over five possible opcodes, and then there's a separate set of add instructions for floating point values. Post-compilation, these instructions are stored in a numerical representation called bytecode. Here are a few examples:

Sample Intel x86 Instructions:

Instruction	Bytecode
add eax, 0x5	83 c0 05
jmp 66fec4 <_main+0x66fec4>	e9 bf fe 66 00
shr edx, 0x3	c1 ea 03

Program Format

Executables are laid out in a particular format that contains information for your operating system on how to run them, as well as extra info for debugging (symbols, function names and boundaries, overall code size, and imports/exports). Code sections, which contain our assembly instructions of interest, may also have data embedded in them (like string values or constant integer values which never change), and functions within these sections are not always laid out in a particular order nor aligned the same way (for instance, with the same amount of padding between them).

Bytecode for "Hello, World!" Program:

```
16128 5548 89E5 4883 EC30 488B 05F9 0000 0048  UH..H..0H. .    H
16144 8B00 4889 45F8 C745 DC00 0000 0089 7DD8  . H.E..E.      .}.
16160 4889 75D0 488B 0565 0000 0048 8945 E048  H.u.H. e    H.E.H
16176 8B05 6200 0000 4889 45E8 8A05 6000 0000  . b    H.E..`
16192 8845 F048 8D75 E048 8D3D 5300 0000 B000  .E.H.u.H.=S     .
16208 E82C 0000 0048 8B05 AC00 0000 488B 0048  .,     H. .    H. H
16224 8B4D F848 39C8 0F85 0800 0000 31C0 4883  .M.H9. .      1.H.
16240 C430 5DC3 E802 0000 000F 0BFF 257F 0000  .0]..       .%
16256 00FF 2589 0000 0000 0000 0000 0000 0000  .%.
16272 0A48 656C 6C6F 2C20 776F 726C 6421 0A0A  Hello, world!
16288 0025 7300 0100 0000 1C00 0000 0000 0000  %s
16304 1C00 0000 0000 0000 1C00 0000 0200 0000
16320 003F 0000 4000 0000 4000 0000 7B3F 0000  ? @     @    {?
16336 0000 0000 4000 0000 0000 0000 0000 0000     @
16352 0000 0000 0300 0000 0C00 0100 1000 0100
16368 0000 0000 0000 0001 0000 0000 0000 0000
16384 0000 0000 0000 1080 0100 0000 0000 1080          .           .
16400 0200 0000 0000 0080 0000 0000 0000 0000          .
16416 0000 0000 0000 0000 0000 0000 0000 0000
```

C Code for "Hello, World!" Program:

C Code
| ```
int main (int argc, char *argv[]) {
    char str[] = "\nHello, world!\n\n";
    printf("%s", str);
    return 0;
}
``` |

The image above shows the bytecode for a simple program in C that prints "Hello, world!". The string "Hello, World!" is embedded between executable code segments, and the corresponding data is highlighted in the image. We can get an idea of which data is instructions and which is the string because we can clearly see the "Hello, world!" on the right-hand side in the ASCII view. But if we were looking at numerical or encrypted data, it wouldn't be so easy to differentiate it from executable bytecode!

Disassembly Methods

Disassemblers typically fall under two types: linear and recursive. Linear disassemblers accept a binary, or a specified area of a binary, and attempt to translate the bytes in it as a single, continuous stream. The problem with this is that, as we saw above, when data is embedded in the middle of a code section, sometimes it can look like valid assembly. Between this and the fact that instructions can vary in length, the disassembler may not be able to tell when it hits a code section, which will throw off the parsing of the overall stream.

Consider this stream of bytes:

```
e8 2c 00 01 00
```

If we start decoding it from the first byte, we get the instruction: "call 0x20002031." But if we begin decoding it from byte "2c," then it will decode as two separate instructions: "sub al, 0x20; add byte ptr [eax], ah."

Recursive disassemblers attempt to fix this problem by following jump, call, and return statements. For instance, a recursive disassembler will continue down a stream of instructions until it hits a jump, in which case it'll calculate the address of the jump, skip to that location, and then continue disassembling from there onward. It sounds reliable enough, but this method also has some pitfalls: jump statements can involve registers, which use runtime information. In this case, the recursive disassembler can't figure out where to jump.
```

```
jmp 0x00ff53 # A disassembler will know to parse bytes at
this address
jmp ecx # A disassembler doesn't know what this jump
location might be!
```

Anti-Disassembly Protections

Large companies seeking to protect their software from reverse engineering and theft, as well as malware authors trying to thwart security analysts, will often strip the earlier-mentioned debug information from their software, including all info about function addresses, symbols, and exports. This means that any potential disassembler has to do a considerably extra amount of work to discover all the possible functions and interpret them correctly, and will not know their names. At the assembly level, there's not always much to indicate the presence of a function. Some code optimizations include function inlining, which takes the bodies of simple functions and embeds them into larger code areas, so you can't always rely on detecting function prologues and epilogues (the typical setup at the beginning and end of functions) to identify them.

To make disassembly even more challenging, companies can inject tweaks into software's bytecode, which will throw off analysis. Here are a few examples:

Junk Code
Random bytes can be inserted into programs to throw off disassemblers. Remember that linear disassemblers will attempt to decode everything in sequence, regardless of the semantics of the instruction; so a software protector might insert some junk bytes after a jump call which are never executed in real time, but will get picked up by the disassembler and throw off the disassembly of everything after it.

Fake Conditional Jumps
Recursive disassemblers will attempt to disassemble both the area after a conditional jump (for example, "jz," which indicates jump if the result of the last operation was zero) as well as the target of the jump. This is pretty reasonable because a conditional branch implies there are two possible paths to take, and either of them could happen during program execution. But anti-disassembly techniques take advantage of this by inserting jumps which are guaranteed to only ever take one branch, and then inserting junk bytes into the other branch to break the flow of disassembly. For example, take a look at the following instructions:

```
sub eax, eax;
jz 0x4110;
```

```
call 0x0;
xor eax,0x5670;
```

The instruction "sub eax, eax" is always equal to zero, and as a result, the following jump ("jz") is always taken. The next line, which is invalid, is never executed when the program runs. But a disassembler won't have this awareness and will include it in its decoder stream. Instead of "xor eax, 0x5670," the code after the jump statement will be listed as "call 0x56703505; add BYTE PTR [eax], al." This misalignment will likely propagate even further down the bytecode.

It's possible to cause decoding issues even without inserting junk instructions. Some anti-disassembly techniques involve breaking the decoding stream by putting a fake conditional jump to an area that is valid, but located a few bytes into an instruction, which causes it and everything following to be decoded incorrectly.

Now, we get to our second question...

WHY DO ANTI-DISASSEMBLY METHODS CAUSE PROBLEMS FOR DISASSEMBLERS, BUT NOT FOR THE OPERATING SYSTEM RUNNING THE PROGRAM?

The short answer to this is that this is because our computers don't disassemble the programs—not in the way we've been thinking of it and discussing it so far. Programs have a saved entry point, called the OEP, that indicates where to begin running its code, and the CPU fetches that instruction, runs it, and then repeats for every following byte. It doesn't know what is coming up in the stream, and if the program has been compiled correctly (which is a fairly reliable assumption in this decade in compiler and software development!), then these potential errors around jumping to the wrong location won't happen. Program execution will simply never encounter the problem areas. The fake jumps always force it to the same path, and it will never hit the junk bytes. Your computer won't attempt to execute data sections as code sections, because program flow simply won't allow it to do so. But disassemblers can't predict what values will be stored in memory or in registers that influence these jumps; they can't possibly predict every correct path to try to guess at them, and they have to make the best educated guess at disassembling as much as possible to give useful information to reverse engineers. For most engineering purposes, disassemblers work well enough—at least most of the program info will be reasonably correct, and reverse engineers can use heavyweight debugging and disassembly software like IDA Pro and Ghidra to manually fix disassembly paths where they need to. You can also use a debugger to execute the code and slowly parse through the correct instructions that way, but this falls less under disassembly and more under dynamic analysis.

FURTHER READING

Branco, R. R., Barbosa, G. N., & Neto, P. D. (2012). Scientific but not academical overview of malware anti-debugging, anti-disassembly and anti-vm technologies. *Black Hat*, *1*(2012), 1–27.

WHY IS IT IMPORTANT TO CLOSE A FILE WHEN YOU'RE FINISHED WITH IT?

The state of the file information in memory, including file contents, any buffers containing file contents, and all the locations for the file, is not finalized until the file is closed.

Buffering is a very common way to avoid expensive system calls to repeatedly read from or write to files. For instance, if you have a program where you are repeatedly writing small numbers out to a file, this written data is often stored in program memory for a time until it hits a certain size, and then is written out. This is much faster than making a system call to store the data on disk every time you write a number. As a result, it's possible for a file to have data in a to-be-written state and not yet written to disk as part of the file. Additionally, files of any decent size are not usually maintained contiguously on disk (or in other words, their contents are split up across different areas of your storage), and so portions of the file may be spread across the drive and the OS needs to determine what data is going where.

Earlier in his career, one of the authors used a 16-bit microprocessor-based machine to run a MS-DOS application for data recording during open-heart surgery. It had to record patient monitoring data and be sure the data was on disk. But there was at the time no command specific for flushing buffers and the only standard recourse was to close a file and then open it again. This would cause the associated buffers to write their contents to disk but it also had an unacceptable overhead in elapsed time—this heart monitoring software had to work continuously to record data as accurately as possible. To get the effect of flushing the buffers without the overhead associated with closing and reopening files, the file handle for the file was duped (i.e., a duplicate was created) and then closed. This caused the buffers to do the write out without the time penalties and overhead associated with closing and opening the file. This is no longer necessary as modern OSes generally have direct commands for flushing buffers, etc., and use much faster chipsets and disk drives than were available at that time.

WHY DOES POST HAPPEN?

POST (Power on Self Test) of hardware ensures the basic hardware and its configuration are functional (and thus any errors that occur are likely occurring higher up the stack, in software).

Computers are a combination of hardware and software. They are notoriously convoluted machines. And once running, errors that occur can be complicated in diagnosis if you haven't first established that the machine itself, the hardware components, are sound and ready for supporting higher computing functions. For example, if you have bad RAM, you want to know that the RAM is bad before you initiate a software process that calls on that RAM. If not, the RAM will cause a failure of the software. The error will apparently occur at the software or application level (on the OSI model). Thus, your search for diagnosing that will likely occur at the software level first, thus ensuring a lot of wasted time and activity.

POST is a firmware-based process that is initiated as the first step for the BIOS (see BIOS) to perform diagnostics on the device's hardware. It does not require an OS to be installed or present. POST does not assume an OS. It does not depend on the presence of an OS to proceed. All testing is done at the system's BIOS level. As a rule, no installed software is called upon until after POST has completed.

It provides a form of proof that the hardware is functioning correctly.

This process includes determining the operational state for the machine:

1. Input device is detected
2. Output or display device is detected
3. Memory onboard meets programmed usage configuration
4. Electrical issues were not encountered on the motherboard (shorts, etc.)

In general, POST proceeds as follows:

1. Confirm functioning system's main memory
2. Initialize BIOS
3. Scan and select which devices are available for booting
4. Verify CPU registers
5. Verify basic components like the clock timer, interrupt controller
6. Pass control to any specialized BIOS extensions

HOW DO PROGRAMS ON THE SAME MACHINE INTERACT WITH EACH OTHER?

Pipes and input/output redirections are the most commonly taught examples of this, and you've probably already been introduced to some output redirection (for instance, if you've saved the output of a terminal command to a file by adding "> output.txt"). But you may also wonder: how can processes interact with each other on their backend? How do programs that monitor or interact with each other do so without requiring command line input and output streams? These types of communications fall under a broad subject called "Inter-Process Communication," or IPC, and there are a few techniques for it.

For one, it's actually possible to use the same sockets interface that your network-related programs (webservers, game servers, and similar programs) use, even though you're working with programs on the same machine that don't need to go out to a router to come back. You can use your "localhost" address to do this, the internal IP address that every machine has (127.0.0.1). Although we typically associate servers with being complex, large programs, it's entirely possible to set up a very small server interface in one program that another on the same machine can connect to or send messages to, and simply set it to only accept connections from the same machine to avoid any accidental interfacing from other computers on the same network.

Another method involves shared memory on the computer system. Every popular OS includes some system calls for programs to register a space in memory that can be accessed by other programs that know the name of that memory. This is somewhat similar to the way you might use a shared whiteboard in an office to leave messages for others, and is similarly fluid in data layout, shape, and content—you can write the data for objects, strings, arrays, and other information in this shared space, as it is just raw memory. For simpler data layouts just requiring strings, programs can use shared files on your filesystem (even just plaintext files), reading and writing data back and forth, although this is a bit of a slower process as disk writes are inherently slow.

Although the sockets technique has built-in mechanisms for handling concurrent connections, shared memory and file writing are a bit trickier, as programs must make sure they don't write conflicting changes with each other, or try to modify the same shared memory all at once. With no guardrails, it's highly feasible that one program might be in the middle of reading data from shared memory when another program decides to update that data, or vice versa. Much in the same way that multithreaded programs use semaphores, locks, and mutexes to protect shared resources from being written to or modified at the same time that they're being used, programs that communicate with each other through shared memory often use identical techniques to make sure there are no collisions.

Finally, most popular operating systems have a feature called a message queue, which is a decades-old backend system interface for delivering messages between processes available on Windows, Linux, and MacOS. This feature includes mechanisms to prevent data conflicts and preserve timestamp information. Although the operating system-based message queues have largely fallen out of popular use with single-system applications today, the message queue concept has been re-implemented in commercial applications for large distributed systems as an effective and robust way to deliver information across multiple servers while preserving message ordering and potential data conflicts.

WHY DO WE HAVE TWO TYPES OF PHYSICAL MEMORY (VOLATILE AND NONVOLATILE)?

There are several types of computer memory, but there are two that are most common: volatile and nonvolatile memory, also called RAM and ROM, or read-write memory and read-only memory. Memory is processed in the CPU and then transmitted to and from secondary memory storage to be kept, processed, and returned.

The BIOS is an example of a ROM-based program. It turns on as soon as the computer is powered up and must be present in order to check that all components of the computer are running properly. When the power is switched off, the BIOS still knows—or remembers—how to boot up and check the data it has access to. Like the BIOS, nonvolatile memory remembers data for when the computer is shut off and stays the same every boot. On the other hand, volatile memory—for example, DRAM—is not stored permanently and loses its data when no power is routed through it.

Information uses memory as temporary, intermittent storage. All changes are performed in the CPU, but the instructions given to the CPU must be stored in memory along with their corresponding data. RAM stores data that the CPU needs to access quickly, and can both read and write data. Whenever you open an application, window, or file on your computer, the data is stored in the RAM (hence, why more RAM usually means better processing performance). It ensures that the data stored is easily accessible when it's needed by the CPU, via a bus (or series of buses) between them. This quick accessibility comes with the trade-off of losing all of that information when the power is shut off.

ROM is nonvolatile, read-only (except when it can be flashed), and usually stores important information that doesn't change often. It is not directly accessed by the CPU and must go through intermediaries before the data can be understood. The BIOS and UEFI are important examples of this, namely boot processes that stay the same. They're mandatory to check that all parts of the computer are functioning correctly, initialize hardware, and organize the boot processes before launching into the operating system or any other operation. Security processes like cryptographic algorithms and keys can also be stored in ROM. These data, and their instructions for the CPU, are not easy to change. Solid State Drives, or SSDs, use a kind of ROM called "flash rom," which is registered and written in blocks of data. This is what allows SSDs to be so fast with data transfers and processing.

FURTHER READINGS

http://learnline.cdu.edu.au/units/hit191/concepts/hardware/reading3.html

https://www.atpinc.com/blog/computer-memory-types-dram-ram-module

https://www.hp.com/us-en/shop/tech-takes/ram-vs-rom?pStoreID=bizclubgold%2Fgb-en%2Fshop%3FpStoreID

https://www.lenovo.com/gb/en/glossary/how-does-a-cpu-work/?orgRef=https%253A%252F%252Fwww.google.com%252F&srsltid=AfmBOooKEvPVOfquTrefy0wQDJt6cI7sDG92D1js8qd0aL9tUw9fgtlK

History

In this section, we cover many of the historical origins of computer jargon and technology. We include things like the background stories for popular, long-running names like "spam" and "Bluetooth," and the QWERTY keyboard.

DOI: 10.1201/9781003519379-2

WHEN SORTING ALPHABETICALLY, WHY DO CAPITAL LETTERS PRINT FIRST?

Capital letters come first because when the computer searches letters in the alphabet, the capital letters appear before the lowercase letters in the American Standard Code for Information Interchange (ASCII) table.

When the computer is instructed to print the alphabet, it starts at the lowest corresponding number for the alphabet's entry in the ASCII table. Capital letters precede lower case letters and so begins with 65 for the capital letter A and from there through the alphabet (66=B, 67=C, etc.).

Please see the answers to the questions "Why do we have the QWERTY keyboard?" and "Why do we have a CAPS LOCK key?"

WHY DO WE USE ASCII LAYOUT?

The ASCII was a US-developed symbol table for managing 8-bit codes that generate teletype-style control codes, simple math symbols, and English alphabet characters, lower and upper case. The standard is defined up to the ASCII code 127. It is implemented in some fashion in every microcomputer today. It was later found that it was possible to significantly extend character sets to 255 to accommodate additional symbols and characters used by other languages.

The English alphabet characters begin late in the ASCII code set as printer head control codes and punctuations such as space, comma, etc., are needed to precede the full alphabet in order to be able to have the controls needed to print the alphabetic characters.

WHY DOES THE ASCII STANDARD START WITH CONTROL CHARACTERS?

Designed when text was physically printed on devices like teletypes, control codes came first in the ASCII sequence because, in order to print characters, you must have control of the printer.

While computers proliferated steadily replacing teletypes, telegraphs, and telegrams, there remained a set of largely printhead-oriented control codes that persisted across the technologies. The bulk of the control codes that evolved for teleprinters allowed a teletypewriter operator to, for example, move a remote print head on a receiving device from the end of a printed line to the beginning of the line, or ring a bell to signal to the remote operator that a teletype message was beginning to come in (or the message transmission had concluded).

Thus, a series of control characters or codes managed print head operation both locally and for the remote or device receiving the communication. Codes changed over the years in usage, often as a function of vendor implementations, such as by DEC or IBM, but the base set stayed within the known parameters. All of these were grandfathered in, or inherited by, the QWERTY keyboard we use today, largely via the adoption of the ASCII. The control codes are the very first encodings in the ASCII character set, for obvious reasons (you have to have control of the printer before you can print something).

Some of the control code implementations, particularly for the original IBM PC, are illustrated in the table below:

Example control codes from the ASCII table:

Using Keyboard	ASCII Code	Command Effect
control+h	08	Backspace (BS)*
control+j	10	Line feed (LF) or move cursor or print head to next line
control+l	12	Form feed (FF) or new page
control+m	13	Return print head to beginning of line (CR)

---

* Control-h (backspace) has a long and conflicted history across OSes and devices. This is partially due to the original teletype print head being unable to move one character space backward (i.e., the destructive backspace). This may be a root reason we have two sorts of backspace today, one destructive of the preceding character (i.e., the usual keyboard meaning) and one non-destructive of the preceding character (now generally used as the back arrow key). All of this has resolved loosely to the backspace key use being the command for the destructive backspace.

These control codes enable the sending operator to, for example, record on a hole-punched tape for transmission, which would then be run completely in batch, including operator-typed errors and correction of those errors.

To see these controls at work, you can, for example, using a terminal window, type control-h on a command line with text to see the backspace happen, or do a redirect of or pipe of, say, control-l to a local printer to tell the printer (usually lpt1:) to go to "top of form" (i.e., in today's single sheet paper feed printers, print a blank sheet of paper).

As is the case with the QWERTY layout for the keyboard, these control codes were carried forward into use in computer keyboards and wholly inherited from teletypewriter usage.

WHY DOES THE ASCII STANDARD HAVE THE ALPHABET STARTING IN THE MIDDLE?

Some sources cite that ASCII A was placed at position 65 for historical compatibility with a UK-originated standard for encoding the English alphabet. Some sources cite expansion of complementary bit offsets and other examples of quick binary math.

The set of codes beyond ASCII 127 represent non-official extensions to support the addition of non-English character sets (i.e., other languages). It is vital to understand that the ASCII extended range is not an official part of the ASCII standard. Thus, you should expect to find a great variety in the wild as to how it is implemented.

The majority of the encoding issues surround internationalization (i18n) of alphabet and character sets as well as the encoding localization (l8n) that occurs in the codes beyond 127. It permits sets of characters for Germanic languages, French, Spanish, and many others but there is no de facto standard for implementation, and so what is implemented varies widely by region, locale, and language.

So, this means if someone sends a message containing an ASCII character code in the extended range, the recipient will not necessarily receive the sender's intended character but rather will see what is defined locally on the recipient's machine for that character code. It will print what corresponds to the local definition for the number, not what the sender intended the receiver to see. This needs to be accounted for and communicated to avoid confusion or misinterpretation.

FURTHER READINGS

https://en.wikipedia.org/wiki/ASCII

https://www.quora.com/Can-the-characters-in-the-extended-ASCII-with-codes-128-255-be-transmitted-reliably-in-single-byte-character-width-when-the-sender-and-the-recipient-use-different-symbols-for-those-characters-in-their-locale-for

https://www.reddit.com/r/explainlikeimfive/comments/mczg8s/eli5_why_do_binary_letters_start_at_65_01000001/?rdt=38444

https://www.shiksha.com/online-courses/articles/difference-between-ascii-and-unicode/

WHY IS IT CALLED BLUETOOTH?

The name draws on the northern European history of a king who united separate realms, just as Bluetooth unites two separate realms (telephony and computing).

"Bluetooth" is reputed to be the nickname of the Danish king, Harald, who united Denmark and Norway in 958. He was so named because he had a tooth that was blue (it is suspected the tooth died and turned bluish).

Bluetooth was designed to bridge the realms of telephony (mobile) and computing hardware (fixed). The purpose was to unite the two realms and support intercommunication.

It was the working title used by the team creating the industry standard. When the specification was finally published, no marketing name had been given to it, so the project name carried over.

The icon for Bluetooth is composed of the king's initials overlapping in native rune script.

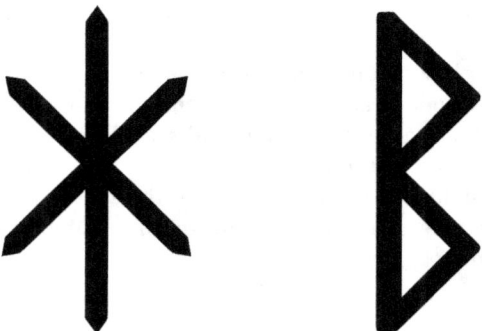

Figure 13.1   Runes for King Harald's initials, which, when overlaid, form the Bluetooth symbol.

FURTHER READING

NPR one bluetooth story: https://one.npr.org/?sharedMediaId=1224766188:1224766190
The org: www.bluetooth.org

WHY IS IT CALLED BOOTSTRAPPING?

In the realm of computer science and technology, bootstrapping has four commonly used, but technically different, meanings.

In general, the term "bootstrapping" originates from an old American phrase, "pull yourself up by your bootstraps." The first recorded use of the phrase was in 1834, when a Montpellier, Vermont resident compared that the continuation of the administration of a democratic government, without practical education to enrich their sense of "rational freedom," was akin to "one [attempting] to lift himself up by pulling at his own boot-straps." In other words, it was impossible.

However, the term changed into a more encouraging meaning at the turn of the 20th century. It became a supposedly empowering way to tell someone to lift themselves by their own hard work. By analogy, when computers still involved the use of punch cards and paper tape, it became customary to have a card possessing a small program that knew how to load larger programs.

Eventually, this evolved into hardwiring the bootloader directly into the computer by soldering it directly to the hardware. This is how the ROM boot was created. Nowadays, bootloaders are normally programmed into non-volatile memory and don't require soldering to modify or repair them.

When a computer starts, it must identify, load, and run the boot code (bootloader) that will then initialize the computer to load the software that creates its capabilities. This boot code is an initial bit of code that loads the larger programs (such as the OS) so that the computer can become useful to a human user.

Bootstrapping also refers to self-compiling compilers, an important facet of programming language development originating from the early ages of language development, where the first instances of compilers had to be constructed from the languages in which they were written. This is when a compiler for a given programming language is written in a different language in order to be understood by the computer, and a new "bootstrapped" version of the compiler is then written in the source language.

This intermediate compiler is used to compile the primary compiler, bootstrapping it, and then this new instance can compile everything else in the source language. Essentially, each iteration of the compiler serves to "pull" the previous compiler up.

When new programming languages are developed, a compiler written in the same language as the source language can be a good example of its sturdiness or overall completeness.

FURTHER READINGS

https://duckduckgo.com/?t=ffab&q=why+do+we+boot+a+computer%3F&atb=v406-1&
    ia=web
https://learn.adafruit.com/bootloader-basics/a-brief-history-of-bootloading
https://medium.com/@akqeel/an-introduction-to-compiler-theory-73395626140a
https://stackoverflow.com/questions/1254542/what-is-bootstrapping
https://web.archive.org/web/20091123154911/
http://www.oopweb.com/Compilers/Documents/Compilers/Volume/cha03s.htm
https://www.howtogeek.com/840484/why-is-turning-a-computer-on-called-booting/
https://www.newspapers.com/article/vermont-watchman-and-state-journal-pulli/
    44074406/
https://www.techopedia.com/definition/3328/bootstrap

WHY IS IT CALLED A BUG?

The conventional wisdom is calling a computer error a bug is based on a moth causing an electrical short in an early, major electronic computing machine.

The first use of the term "bug" in the context of computing—in the sense that you can have a "bug" in something and also be able to "debug" it—is the archetypal story around early computer pioneer Rear Admiral Grace Hopper and her team working on an early mainframe, the Harvard Mark I.

Many authoritative sources cite this story:

The early days of computing involved massive arrays of wires and tubes, and space. The computer hardware—later always the case with mainframes—filled up rooms. Much of the circuitry used for the technology was exposed to open air. The team had a system failure or short circuit and while investigating the source of their trouble, one of the team found the carcass of a moth had caused a short. They stated, upon discovery of the moth, that there's a confirmed bug in it.

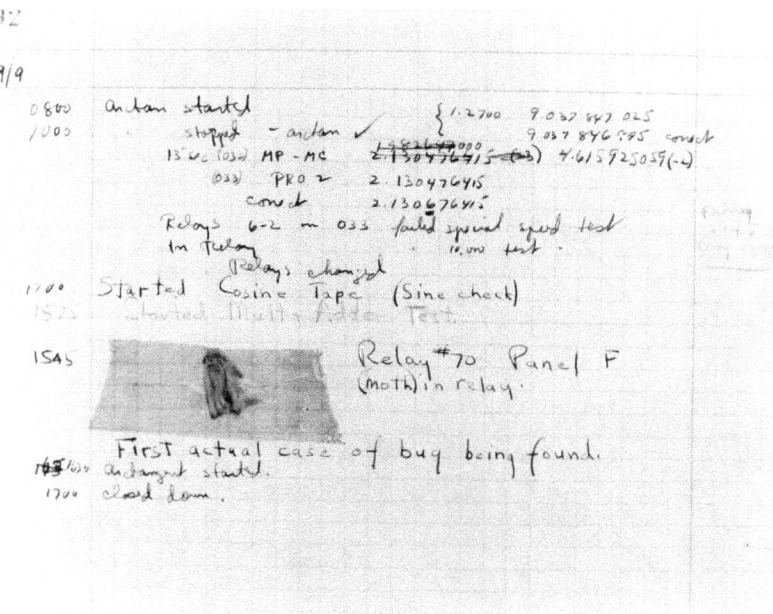

This file is a work of a sailor or employee of the US Navy, taken or made as part of that person's official duties. As a work of the US federal government, it is in the public domain in the United States.'

https://www.history.navy.mil/our-collections/photography/numeri-cal-list-of-images/nhhc-series/nh-series/NH-96000/NH-96566-KN.html

It is not the first reference to a literal bug in a system causing the system or a feature of the system to fail. There are notes by Thomas Edison when he was improving the telephone transmitter that refer to removing bugs from his prototypes. A bug would enter his telephone prototype in his lab, die there, and then the carcass would cause the machine to fail.

Further historical examples of the use of such phrases concerning bugs and failures—for example, a "fly in the ointment" being of long English usage—can be found in print as early as 1707 (author John Norris). This would indicate bugs have a long-standing history of disturbing human inventions.

There is even a much older reference to our problem with bugs breaking things in Ecclesiastes 10:1: "dead flies give perfume a bad smell."

FURTHER READINGS

https://www.howtogeek.com/726020/what-is-a-computer-bug-and-where-did-the-term-come-from/
https://www.phrases.org.uk/meanings/fly-in-the-ointment.html

WHY DO KEYBOARDS HAVE A CAPS LOCK KEY?

Today, the caps lock key is inherent in the generally adopted QWERTY design of the keyboard and provides backward compatibility—albeit slightly dysfunctionally—to the original typewriter and teletype keyboards (which printed in capital letters only).

All typewriters originally produced only capital letters, resolving the fixed baseline problem (have no characters that go below the typed line) that occurs with mechanical type-based machines.

As mentioned in the QWERTY entry, when new typewriter innovations allowed for lowercase characters, the shift key was implemented, so typists could switch between cases. However, the shift key—a physical piece of metal which ran underneath the type bar—became too fatiguing to hold down for long stretches of time. Thus, the shift lock key was conceived. Shift lock was first seen in the initial Remington Portable typewriter model, released in 1920.

However, the shift lock remained tiresome for a different reason: when it was engaged, it would not only capitalize letters, but also change numbers to special characters as well.

Purportedly, in the mid-1960s, a Bell Laboratories telephone engineer saw how frustrating the shift lock key could be. His boss's secretary would type and then accidentally, using the SHIFT LOCK key, turn numbers into special characters.

She would swear loudly and throw away the page of work she'd done so far. So, he began experimenting with typewriters, and in 1971 was granted a patent for his "cap" key idea, a button which would only capitalize the alphabetic characters.

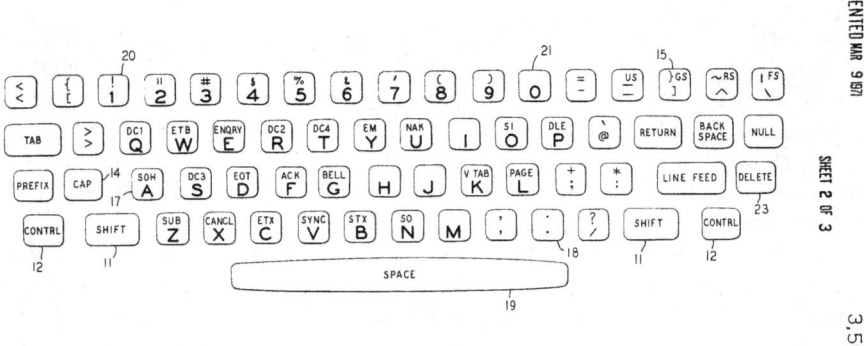

FURTHER READINGS

https://medium.com/forwardtick/its-time-for-caps-lock-to-die-81c9eaa4dfa7
https://patents.google.com/patent/US3569991
https://site.xavier.edu/polt/typewriters/rem-portables.htm#junior1
https://www.splinter.com/i-turned-caps-lock-on-for-a-week-and-everyone-hated-it-1793845143

WHY DOES PRESSING THE KEY COMBINATION OF CONTROL+G RING A BELL?

It provides backward compatibility with older style input devices, such as teletype printers, which communicated by phone lines and needed a way to signal to a human operator that a message being printed on their local machine is beginning (or has ended). It was keyed into a remote machine by the remote operator sending the message.

On telegraph and teletype machines, it was not only possible to send text messages to other teletype machines, but also to control the printing at the remote teletype printer. There were various control codes used for managing the printer head (new line, return to beginning of line, etc.), paper feed, and the announcement of the beginning of a message (or of the end)... by ringing a bell.

All of these control codes are included in the ASCII character set standard. These days, the cursor usually plays the role of print head.

Example control codes from the ASCII table:

Keyboard	ASCII Code	Command Effect
control+g	07	Ring bell (BL)
control+h	08	backspace (BS)
control+j	10	line feed (LF) or move print head to next line
control+l	12	form feed (FF) or new page
control+m	13	Return print head to beginning of line (CR)

WHY DO MONOSPACED OR FIXED FONTS EXIST?

Monospace, or fixed-width, fonts were originally the only type available due to the mechanical technology (keys, stamps, etc.) used to print characters on paper by typewriters or teletype printers.

With the typewriter, and then later the teletype, a mechanical printer-type communication device which utilized serial messaging, monospace font was a side effect of how characters were mechanically printed on paper (refer to "Why is the terminal labeled TTY?"). These devices used impact printing to generate characters. Impact printing means every character must be physically the same height and width so that the striking character head has the same baseline.

A desired letter key is struck, causing a lever holding the corresponding character stamp to be propelled forward onto an ink ribbon placed over a piece of paper held against a platen backing. These character stamps had to be the same width and height for the levers as for the printing. This had the side effect of a line length total of 79 characters for a standard printed line. And that 79-character length "column width" standard was adopted into such things as "green screen" terminals.

It was only later, with graphical user interfaces, did it become possible to have variable-width font and variable font printing and display. This includes visual adjustments such as kerning and other typographic layout techniques.

However, programmers, IDEs, and terminal interfaces still use monospaced fonts to this day. The common sentiment behind this is that it's easier to align, read, and debug code if all the characters take up uniform space. Monospaced fonts draw attention to individual characters rather than whole strings of text—so if you're looking for a specific syntax error or misspelling, your eyes can hook onto a specific character easier than if you were using a variable-spaced font (see below).

Monospace fonts and variable-width fonts examples

```
The quick brown fox      The quick brown fox jumps over the
jumps over the lazy      lazy dog.
dog.
```

Monospace fonts focus more on individual characters.

```
System.out.println ("Hello World!");
System.out.printnl ("Hello World!");
```

Since variable-width fonts focus less on individual characters, it's tougher to spot things like syntax errors.

FURTHER READINGS

https://typetype.org/blog/monospaced-fonts-in-design-and-programming/
https://www.sitepoint.com/the-anatomy-of-a-letterform/

WHY DO WE USE THE QWERTY KEYBOARD?

The QWERTY keyboard (so named because of the sequence of the five top left row keys) was designed to deliberately slow the typist down.

Typewriters came into existence in the 1800s and initially only allowed for capital letters. With the introduction of the shift key, later models allowed for both uppercase and lowercase. Early typewriters were cumbersome and generally not faster than writing by hand. This led over time to improvements in typewriter technology and to faster typists.

Typewriters then developed a new problem: they were subject to frequent jamming by typists fingering too quickly. This would cause jamming that then caused mistakes or freezing of the typewriter action. This would result in the typist having to reset the device. This did not engender a good user experience.

In 1867, American serial typewriter innovator Christopher Sholes invented the QWERTY key layout to reduce jammed keys (versus a straight alphabetical layout was used) because it took longer to type on a QWERTY layout. That is, this layout became widely adopted because it slowed the typist down.

Typewriters became immensely popular, spreading the QWERTY keyboard far and wide. It was further embraced when teletype machines—replacing the specialty Morse code system—proliferated globally (as they could be used by anyone, and not just by folks trained in Morse). Thanks to decades of experience with the QWERTY keyboard and generations of trained typists already familiar with the layout, the QWERTY keyboard easily extended to other devices needing keyboards, such as the teletype and then computers.

It is sometimes suggested that the Dvorak keyboard—invented in 1936 as a reaction and with the intention to be a more rational alternative to the phenomenal success of the QWERTY keyboard—is superior in speed and ergonomics but with the exception of a handful of admirers, it has never seen serious adoption.

WHY IS THE SAVE ICON THE WAY IT IS?

The save icon image is a representation of a portable storage device called a diskette, or floppy disk—originally starting at 8″ and then gradually getting smaller, down to 3.5″—the most commonly used storage media for personal and work computers from the 1970s to the 1990s. This was before higher density and storage media such as CDs, USB drives, and SD cards became prevalent.

Typically, visual icons are chosen to represent the function they're intended for—the trash shortcut is a trash can or recycle bin, buttons to upload or move data contain arrows, the filesystem explorer icon usually contains a folder, and so on. Save icons are generally representations of storage media, but the most common one seen on modern OSes and user applications isn't quite as obvious or well-known. It is of the data diskette, an extremely popular form of physical and portable storage in the late 1900s, and a precursor to things like CDs and USB drives. It likely became the widespread symbol of saving data at the time because people have actually handled them directly, whereas other types of storage, like RAM, or hard disk drives (HDDs), are internal, and not normally visible to users.

In the early days of microcomputers, various icons were tested as GUIs (Graphical User Interface) became popular (e.g., a cylinder, a simple oblong shape, a representation of an internal hard disk drive, etc.), and over time, the floppy disk became the most readily understood. Although the floppy disk has been largely obsolete for years, it has remained a pervasive representation for data storage. With things like cloud storage and remote file synchronization, who knows what the next conceptualization of data saving might be.

WHY IS JUNK, VOLUMINOUS, OR UNSOLICITED EMAIL AND MESSAGING CALLED SPAM?

The traditional story about the origin of the use of the word spam—as in unwanted or excessive messaging, email, or postings—is generally credited to a skit on the British TV series Monty Python's Flying Circus. It was first broadcast in 1970 on the BBC. During the skit, the word "spam" is used excessively. (On a side note, Monty Python humor has a long history of popularity in CS circles and early groundbreakers in the field were clearly all Monty Python enthusiasts.)

In the opening of the skit, two main characters—presumably husband and wife—descend on chairs from above the camera frame into a cafeteria next to a table of Vikings. They are approached by a waitress who proceeds to tell them all of the offerings available... almost every one of which includes spam as a main ingredient.

Eventually, the skit itself eventually becomes overwhelmed by the use of the word "spam" ("spammed") and the unrelenting references to spam. It then has a non-English character whose translation dictionary is overloaded with the use of "spam"—who is then forcibly removed by a policeman—and then cuts to "A Historian" (sic) discussing Viking victories involving spam, which ends with everyone back in the dinner chanting "spam." While the skit is about 4 minutes, there are 67 mentions of spam in the opening dinner scene portion alone. This includes the incongruous Vikings in the dinner scene, chanting "spam!" at every opportunity and particularly in overwhelming cacophony toward the end of the skit.

FURTHER READINGS

The Monty Python skit video: https://www.youtube.com/watch?v=ycKNt0MhTkk
Historical references: https://cacm.acm.org/research/the-history-of-digital-spam/
https://www.edn.com/1st-spam-email-is-sent-may-3-1978/
https://cacm.acm.org/research/the-history-of-digital-spam/

WHY IS THE TERMINAL LABELED TTY?

In 1928, The Teletype Corporation created a remote serial printer they called the teletype and eventually that name became the generic term for the hardware. It was abbreviated "TTY." For many years, the teletype was the de facto connection and terminal interface with computers and was eventually virtualized for use in every computer today to issue commands directly to the OS kernel.

The history of the TTY is a mess of satisfying prior needs and retaining backward compatibility to prior standards applied to popular but often very dissimilar underlying technologies.

TTY is the abbreviation for the teletype (usually a keyboard connected to a printer with a roll of 8.5-inch paper feed and an adjacent "ticker tape" strip printer, generally for creating batch input and output streams). TTY can refer to either physical and/or virtual forms. As mentioned above, physical teletypes are printer-based systems with keyboards used for serial communication to other teletypes or to mainframes, and virtual teletypes are emulations of teletype-based serial connections generally made directly into the OS kernel.

The original teletype evolved as a serial communication device between keyboard terminals with built-in printers, in the early days replacing Morse code as the standard general communication method, eventually becoming a serial tool for communicating with computers, especially the first computer, the Mainframe. The printed tape was used to record instructions, which were then loaded via the teletype into the mainframe for processing (a process referred to as "batch" or batching).

The PC or laptop console (or the command-line interface aka "CLI") is a TTY because it is emulating a (serial) teletype connection to the operating system kernel. The terminal window is in turn behaving as a device that only manages serial input and output (originally, connection to another teletype, and later, to a mainframe computer). This is why you often see odd responses or frequently have to provide filters to TTY commands, because the machine is operating as if it were a sharing resource across multiple users or devices, as would be the case with terminals accessing a mainframe.

FURTHER READINGS

https://www.howtogeek.com/428174/what-is-a-tty-on-linux-and-how-to-use-the-tty
    -command/
https://www.howtogeek.com/727213/what-are-teletypes-and-why-were-they-used-with
    -computers/
https://www.linusakesson.net/programming/tty/

WHAT WAS Y2K AND WHY WAS IT A BIG DEAL?

The "Year 2000 Problem" is the theory that there would be widespread hardware and software malfunctions at the turn of the century because of a quirk in numerical year representation.

In the 1960s, when computers had comparatively minuscule memory and storage capacity, many programmers stored the calendar year in two digits instead of four. There are multiple (and often synonymous) theories for this: that it would cost upward of five times the memory storage to represent a four-digit year, or that programmers thought the code for their machines would be scrapped in a few decades' time, or that nobody would use their programs so far in the future.

However, going into the 1980s and 1990s, older code was still widely used in many computer systems. Many computer scientists worried that these computers would be unable to differentiate between 2000 and 1900 ("00"). When the new millennium struck, the date change could result in system-wide malfunctions from time inaccuracies or data corruption.

In a report written for the US Senate by the Committee on Government Reform and Oversight, it was theorized that vulnerable systems could be the same ones responsible for hospital systems, bank records, transportation software, stock market algorithms, and all sorts of digital communication services. More pressingly, microchips (clocks in embedded systems) were liable to malfunction. Anything with an internal clock could be susceptible to failure, from microwaves to traffic lights, plane flight timetables to elevators.

So, with the supposed threat of widespread computer malfunction on the horizon, the Y2K problem was born. A preventative solution was drafted for the US House of Congress in 1996, with emphasis on disaster and panic prevention.

> The point of calling for such urgency is not to trigger panic, but in fact to avoid panic. If this problem does not receive the attention it demands during the next 6 to 9 months, and if we allow the date change to approach without knowing our vulnerability, panic will be the inevitable result. The only way to avoid this is to act now. The President must sound the alarm and address to the Nation now in order to avoid panic later.
>
> Dan Burton (October 26, 1998).

Plans were drafted for various government wings to check, update, and verify each individual computer in use. Public warnings were issued not just in the United States, but also in various European countries, Australia, and parts of Asia. To a lesser extent, countries in South America and Africa were also notified. Each country could be drastically different in policy implementation. Where France, Australia, and the United Kingdom were efficient in dealing with the problem, other countries like Italy or parts of the Balkans responded with little urgency.

Nearly everybody working in IT at the time (at least in the United States) dealt with the issue of Y2K. Their work largely went unnoticed by the general public. Approximately 300 billion USD was spent on Y2K prevention globally, with about half being in the United States alone, according to a study by Gartner.[1] A majority of these fixes were done by "windowing" the systems, or by playing a logical trick on the computer's memory and extending the window of time forward a few decades (commonly 30 years). These fixes essentially applied a "band-aid" to the issue of Y2K with the hopes that the technology would be phased out, replaced, or that future programmers would eventually fix the problem.

Y2K speculation didn't end just on December 31, 1999 or January 1, 2000, however. In one instance, there were speculative concerns that the date 9/9/1999 would cause computer failure because programmers indicated a program termination with a string of 9's. The year 2000 was also a leap year—the first of its kind in 400 years.

Due to how the Gregorian calendar works, February adds a 29th day to its calendar month in all years divisible by four. The exception is when the year is divisible by 100 (so years like 1900 or 1700 were not leap years) unless the year is also divisible by 400. For the first time since 1600, there was a leap year at the turn of the century. There was a concern that there would be an additional day in the year, unrecognized by computer systems, because some of the programmers behind their clocks assumed the year wouldn't "leap forward." A government agency in the United States was set up for monitoring this leap year problem, with much less urgency than Y2K. No major issues were cited as a result of the potential fault.

There is still much debate on whether the preparation program for Y2K was worthwhile. Some experts argue that the Y2K crisis was based on speculative reasoning or that many systems continued past the new year without issues. At the same time, in the failures that did occur (namely with small companies), entire databases, financial records, and patient histories were altered or removed by the system. If these issues were replicated in essential governmental, airline, or large medical networks, the potential catastrophe could have been disastrous—it comes down to a question of hardware and software working together.

The biggest Y2K-like problem in the future will be on March 19, 2038, at 3:14:07 UTC. Coming back to microprocessors and chips, 32-bit systems have a maximum integer ceiling of 2,147,483,647. Microchips log time in seconds, at least for the user's display. If you break down the time passed since January 1, 1970, into seconds, that specific date and hour in 2038 will be the total seconds elapsed since Unix computers began counting time. Thankfully, most computers nowadays are 64-bit systems, and their ceiling of 9 quintillion far surpasses that of 2.147 billion. When March 19, 2038 comes around, however, any 32-bit Unix systems still in operation could potentially get fossilized in time.

An interesting example of this actually came from Psy's "Gangnam Style" music video on YouTube. At the time of the video being posted, YouTube's view counter was still 32-bit. So, when Gangnam Style became viral and hit exactly 2,147,483,647 views, the counter number came to a halt until the programmers at YouTube were able to upgrade the counting system to 64-bit. It became a viral example of what will likely happen to 32-bit Unix systems in the year 2038, as well as a valuable reason for why the upgrade to 64-bit systems is so important.

FURTHER READINGS

http://www.lieberbiber.de/2017/03/14/a-look-at-the-year-20362038-problems-and-time-proofness-in-various-systems/

https://archive.nytimes.com/www.nytimes.com/library/tech/99/07/biztech/articles/05gart.html

https://archive.ph/aAfEa

https://smartermsp.com/tech-time-warp-the-leap-year-bug-y2ks-lesser-known-counterpart/

https://www.congress.gov/105/crpt/hrpt827/CRPT-105hrpt827.pdf

https://www.hpcwire.com/1999/03/19/common-y2k-quick-fix-last-decades/

https://www.scientificamerican.com/article/what-are-the-main-problem/

https://www.si.edu/spotlight/y2k

https://www.theguardian.com/technology/2014/dec/17/is-the-year-2038-problem-the-new-y2k-bug

NOTE

1 Gartner, an Early Y2K Analyst, Dominates the Niche. https://www.si.edu/spotlight/y2k.

Infrastructure

This section more generally covers questions around the computing infrastructure we interact with day to day as users and as computing professionals. It explains some of the hardware and software design elements in common, everyday software and hardware we use.

DOI: 10.1201/9781003519379-3

WHY DO LANDLINE PHONES HAVE DIAL TONES AND CELLPHONES DO NOT?

Traditionally, cell phones and landlines are using two fundamentally different telephony technologies. One is request-based and the other involves active listening. Cell phones use a wireless call request (using distributed local cell towers operated by various carriers) and so have SEND and END buttons to initiate and terminate calls. Landlines are using an always-on line (using a heartbeat line to a central serving authority) wired communication system and are in listen mode, signified by the dial tone.

Calls on cell phones do not begin until the phone requests a connection through a cell tower. Landlines have an always-on system—hence the dial tone when you lift the receiver, which informs you in general that the system is working and awaiting a phone number. You do not know if the cell phone call will work until you initiate it (although the strength of the cell tower signal—the number of "bars" locally on the phone—will give you a good estimation of the likelihood of the call being successful, if the number you enter has a valid destination).

With a landline, the service is always listening. When the receiver is lifted, the line is activated and this is signified by the sound of the dial tone. With cell technology, the system does not begin listening to a call or call request until the phone initiates a call to connect with the local cell tower.

Interesting history: Because landlines are always listening for a tone when there is dial tone, they were once susceptible to manipulation or phone hacking ("phreaking"). Not widely known now, but the co-founders of Apple Computer, Steve Jobs and Steve Wozniak, created and sold what was at the time called a "blue box" device, which emulated the service tones into dial tone necessary to make phone calls domestically and internationally. The story is that Wozniak finally stopped using it after posing as the US Secretary of State Henry Kissinger and dialing the Vatican in Rome, convincing the staff to awaken the Pope for an urgent conversation.

FURTHER READINGS

https://en.wikipedia.org/wiki/Blue_box
https://www.thehenryford.org/collections-and-research/digital-collections/artifact/452666#slide=gs-427299

HOW IS IT POSSIBLE THAT FORENSIC ANALYSTS CAN RECOVER FILES AFTER THEY'VE BEEN DELETED?

It's a common misunderstanding that files are permanently, immediately destroyed when deleted from a filesystem. Even if you empty your trash bin right after you delete a file, there's a chance it can still be recovered—this is how forensic investigators can recover data from a suspect's laptop. It's not foolproof, however, and has to be done quickly after the file is deleted.

The primary reason is that the data inside your files isn't actually wiped out immediately when you delete something. Rather, just the metadata that helps your OS keep track of the file is deleted. Let's say you have a simple text file containing a few paragraphs located in your Documents folder. When you view the Documents folder and its contents, your operating system knows about what's located in it because it keeps a tally of files with their names, sizes, filesystem locations, and where their associated data is stored on your computer's physical hard drive. This info is essential because your Documents files, for example, aren't necessarily located next to each other on disk; they may be broken up into separate areas or stored wherever there's the most space available(this is a whole separate science, called "file allocation," if you want to read more about the theory behind placing data). On Linux, this metadata is known as inode information; on Windows, this info is stored in the MFT or "Master File Table."

When you delete a file, just the metadata to that file is deleted, but the underlying data remains, and the memory addresses it occupied are marked as "unallocated" or unused. The reason is simply that fully erasing data is extra work for the operating system that it's usually just considered unnecessary; this would require writing other data, for instance "00," over every byte of space in the file before deleting it. SSDs, or solid state drives, which are a popular hard drive in modern laptops and desktops, have a limited number of times they can update data before wearing out, and so wiping out every byte of deleted data would age them prematurely (although newer SSDs are so resilient, it would still take a significant amount of time for this to become a problem).

Forensic analysts and file recovery services use this information to scan all unallocated memory for artifacts. Common file formats including word docs, JPG, PDFs, and text files have characteristic structures at the byte level, and analysts will scan for these to determine if a particular stretch of unallocated memory contains the data for a file. To get an idea of what some of this data might look like, you can even examine it yourself without having to delete anything or scan your filesystem: download a hex editor, or try a site like https://hexed.it/, and open a JPG or PDF. Most of it will look like gibberish, as the graphical components are designed to be parsed by an image viewer, but you might see some identifying bits like "PDF" and "JPG" translated from some of the bytes. Here, we include an example:

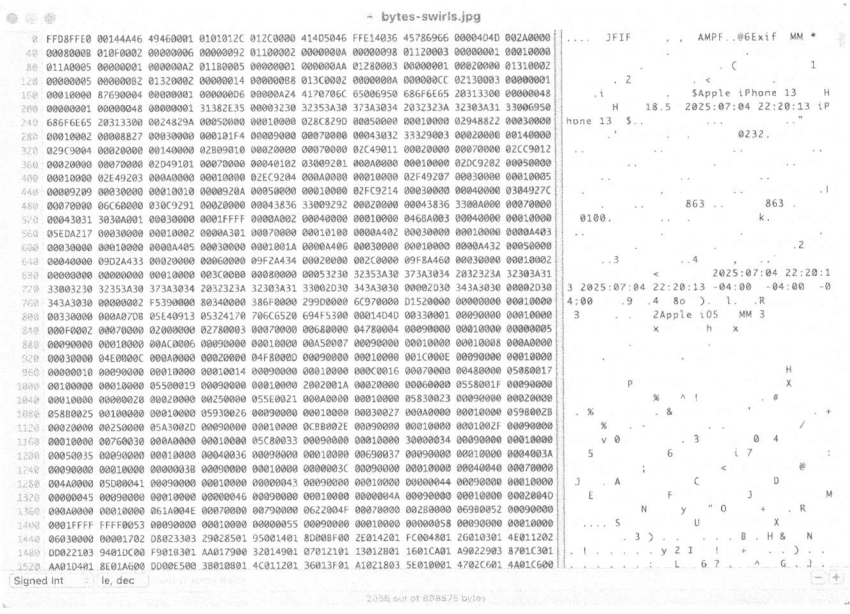

In the upper photo is a JPG image. In the lower photo is the same file viewed through a hex viewer, where you can see some identifying traits like "JFIF" and the type of camera that was used to take the image. Analysts use software that can reconstruct all of this hex back into a viewable file, as well as retrieve info about the device used to take the photo, time, and location.

FURTHER READING

https://forensics.wiki/file_carving/

WHY FRONT-END VS BACKEND INFRASTRUCTURES?

Front-end and backend infrastructures are needed to accomplish entirely different jobs with different objectives requiring different technologies and expertise, one addressing windows and presentations the user operates, and the other in databases and APIs where computers primarily operate.

The technical requirements and major differences between front-end and backend infrastructure have often created different levels and specialties of developers, each with their own domain challenges, skills, and expertise.

Examples of issues at the front-end and backend

Front-end issues	Web page display
	Page design and layout
	Functionality
	UI/UX
	Authentication
Backend issues	Database management
	APIs
	Interconnectivity
	infrastructure

While it is possible to find developers who fully practice both, they are rare and will likely become rarer over time, except for the most trivial of applications, as the core skillsets and tools continue to diverge significantly. The developer who can provide both front-end and backend development skills is called a "full-stack" developer. While this is often a very desired position to fill in the industry, they tend to be expensive hires.

It is also important to understand that the UI and UX timescales are very different for these two domains due to what they do and interact with. This reflects some of the differences in technologies and skills the developers need to excel in.

User-facing UX/UI behavior is generally measured in human-facing response times, like seconds. Backend work is usually compute-facing technology with database servers, APIs, and transaction servers. These are measured in computer response times, measured in ms or ns (depending on the type of transaction or data involved).

FURTHER READING

https://medium.com/js-dojo/modern-frontend-architecture-101-f9c88c20ea20

WHY ARE THERE SO MANY DIGITAL GRAPHIC FORMATS?

For most online and print images use, there are two common graphic types to pick from: bitmap and vector. One is used for screen presentations (bitmaps are derived from pixels) and the other for mathematical construction of imagery (vectors are derived from calculations).

Every bitmap graphic has the ideal resolution for presenting its bitmap image. This is set by the pixel resolution of the bitmap itself—how fine the pixels and color composition are. And that's the perfect size for the presentation of the bitmap image. This varies by type of bitmap graphic. Bitmaps are resolution-dependent. So, you need to know the bitmap's best use.

Vector graphics are not resolution-dependent as they generally use mathematics —usually descriptions in the form of vectors—to define their image. The files that contain them are thus usually many times larger than bitmap files. However, they may be used to make a presentation at any resolution (and this is generally why the publishing industry requires them).

In general, graphic formats come in and out of popularity as a function of their application and usefulness. With few exceptions, formats give way to newer formats that are more compact, produce better resolution, or load more quickly because of their reduced file size. Long-standing or entrenched favorites do exist. These include JPG for web images, RAW for professional-grade photographs, EPS for displays and printed publications, and PDF for documents and use of signatures.

Digital graphic formats have advanced significantly in quality and usability over the years. Much of this is due to improved algorithmic methods of managing data and storage. Additionally, file formats and image types surge in popularity as a function of the platform they are primarily used on. Some formats, which are lower in resolution, have come to dominate web platforms because they are relatively small and readily transferred over an internet transport layer such as HTTP. Others have come to dominate in high-resolution applications (such as printing photographs and glossy brochures) because they are rich in resolution, not transferred frequently, and are stored on large-capacity storage devices.

This leaves a range of format types which change over time, governed by a few basic principles. The features of the various graphic types interconnect and interact in important ways:

- Size: The larger the file, the more detail it can contain for generating the graphic. Additionally, the larger the file, the longer it will take to move the file around or transfer the file to other mediums, whether by tether, wirelessly, or over a network.

- Resolution: The lower the resolution of an image, the smaller the file that contains the data and thus the easier and faster to transfer.
- Scalability: The trade-off here is that, as a smaller image is expanded or made larger, it will lose clarity and resolution and become spotty and blurry, if not downright unrecognizable and/or unusable.

The essence of the selection process for the best image rendering format is to pick a format that ensures the data is there in the file to handle adjusting the image to its intended use (for example, "blowing the image up") across a range of targets including webpage, printed brochure, or billboard.

Another changing-over-time detail has been the dramatic improvement in internet connection speeds over the last decade. This has given rise to the impression that data (and hence file) size is no longer a relevant issue but this is untrue. While the average webpage has grown from kilobytes into megabytes in size in the last decade, a truly dense or composite photographic image will exist as data that will be an order of magnitude (or more!) larger than that webpage. This size and detail will have a profoundly negative effect on the user experience for a website (due to the lengthy loading time of the image and/or the overwhelmingly large display size) but very positive for a photographer creating a photographic portfolio (due to the high quality of the image).

Understanding where the file format came from—what it was used for—is a basic question that divides most graphic file formats into two types: screen capture or print capture. The basic question is: is this file format originally for display-screen-optimized or image-optimized? The original screen-based graphic file formats were generally derived from raster displays, where the images were rendered with laser sweeps creating pixels, unlike modern displays, where pixels are rendered individually or by LED, for example. Raster refers to the raster sweep of the CRT guns (in red, green, and blue colors, hence the acronym term RGB and there is also CMYK and others) used to fill in the display of a CRT display with a fixed pixel placement. So, this type of file format or image is referred to as "raster" imagery. Bitmap formats tend to be mapped in an RGB-optimized manner. This technology is similar to older TV sets and has long been used for driving monitor displays (characterized by a prevailing ratio between the size factor of the width of the display as directly related to the depth of the tube and thus the case housing the tube).

Vectorized formats are necessary for print images and for large-scale images such as posters or billboards. So often, with these sorts of images to view them on a computer requires a special program to display the image.

FURTHER READINGS

Adobe on graphic files: https://www.adobe.com/acrobat/hub/guide-to-image-file-formats.html

Hi lo res, vector, etc.: https://blog.hubspot.com/insiders/different-types-of-image-files

https://en.wikipedia.org/wiki/Jumbotron

https://kinsta.com/blog/image-file-types/

https://www.geeksforgeeks.org/image-formats/

https://www.graphicdet.com/blog/understanding-file-types/

https://www.sciencefacts.net/cathode-ray-tube-crt.html

Networking and Security

A handy look explaining some common phenomena and conventions encountered when studying networking and security-related concepts, such as web cookies, public key infrastructure (PKI) infrastructure, latency, and device identifiers.

DOI: 10.1201/9781003519379-4

HOW DO WEB COOKIES WORK?

Web cookies, also called HTTP or internet cookies, are small packets of data that get sent from a web server to a user's browser. When you visit a website, you send a request to the website's server in order to view its page content. The web server sends a response back to your browser, which tells your browser to store the cookie (data packet) for its website. This cookie is stored in a dedicated cookie file in your browser directory. Whenever you go back to the website, your browser sends the cookie back to the server.

Servers don't necessarily store cookie information. Instead, they keep track of visitors by storing what are called "session IDs" (SIDs). When a user receives a cookie from a site, the session the user instantiates is logged in the site's web server: this normally contains a record of user activity and certain identifying characteristics in an alpha numeric header. Web servers match SIDs with session cookies to remember certain user activity. Thus even when a user leaves a site and comes back, the special session ID will match the locally stored data on their device and often pick up where they left off.

Cookies have a varied number of uses: depending on the type of cookie, they manage all sorts of functions. They can be used to:

- store your login information;
- build things like shopping carts;
- create lists which require an ability to remember your items across pages;
- customize and refine your user experience (UX) on websites;
- more broadly, track your preferred content across the Internet to be used for customizing advertising to you.

Non-Persistent or Session Cookies

Non-persistent cookies, as the name implies, expire once a user ends a browsing session, either when they close the target website or when they close their browser. Normally, session cookies remember a user's actions or preferences when a user is actively on a website, and can assist in site functions like shopping carts or wish lists.

First-party cookies are a type of session cookie, and are set by the website you are directly visiting, not by sponsors or ad services. They commonly store website preferences (like language choice, theme, region, etc.)

Persistent Cookies

Persistent cookies are stored on your computer for some set span of time—this could be seconds, days, or even years. While they have a predetermined expiration date, this can vary widely depending on the service the cookies are used for. Services such as browsing habit tracking, as well as general cross-site cookies that can personalize the types of ads you receive, fall under this category.

Third-Party Cookies

Third-party cookies are a type of persistent cookie set by third-party services, like advertisers. These are normally in the form of embedded media, such as video or image advertisements, and are primarily used to track your behavior across websites.

When you visit a website, you load ads from third-party services, which set cookies on your computer. As you move on and visit other sites that contain ads from those same third-party services, the cookies planted on your computer by the ads allow your activity to be tracked across the Internet.

Supercookies

Supercookies, while not functionally web cookies, are named such because of their accelerated ability to track user activity across the Internet. The name "supercookie" is a general term for bits of tracking code that keep tabs on user activity in far more pervasive and consistent ways than web cookies. They can take various forms, from data simply being stored outside of your cookies folder in your computer, all the way to unique data headers that are injected directly into the data you request when you visit a website.

Supercookies can be stored in various places in order to avoid detection, deletion, or blocking, like in your cache folder or other places in your browser's file system. Alternatively, they can also utilize certain http traffic headers, which are communicated every time you open a website or request information from an http-encoded site. By injecting headers—somewhat like tracking tags—into traffic requests, companies can completely subvert normal cookie allowances and form far more comprehensive data profiles on users.

In 2016, Verizon and various other mobile data carriers were found guilty of using supercookies to track user's mobile data habits. They would track anything from gender, age, interests, and hobbies, down to email and physical addresses, URLs visited, and apps used.

Now obsolete everywhere outside of mainland China, a variation of supercookie was something called Flash cookies, utilized by Adobe's multimedia Flash plugin. In contrast to supercookies being injected into network traffic, these were bits of data stored on your computer outside of the browser file system, which could not be controlled or deleted using native browser privacy controls.

Supercookies had a greater storage capacity compared to web cookies, could track internet behavior across different browsers on the same device, and had a particularly ominous subtype called a "zombie cookie"—a bit of Flash code that would regenerate any deleted web cookies in a user's cookie folder. Zombie cookies were often utilized by companies and ad networks to collect user data and create user browsing profiles.

The name for browser cookies was coined by the inventor of cookies, Lou Montulli, and is derived from the term "magic cookie." Much like a fortune cookie, the name is meant to represent data stored inside a file, which is only known by the software transmitting it, not necessarily the user it affects.

FURTHER READINGS

https://blog.mozilla.org/en/internet-culture/mozilla-explains-cookies-and-supercookies/
https://proprivacy.com/guides/super-cookies-flash-cookies
https://security.stackexchange.com/questions/225620/is-cookie-information-stored-on-the-server-side
https://time.com/3695826/supercookies-cookies-verizon-att/
https://www.cookieyes.com/blog/internet-cookies/
https://www.forbes.com/sites/ygrauer/2016/03/14/fcc-slapped-verizon-with-a-1-35m-fine-but-supercookies-remain-a-problem-around-the-world/
https://www.hp.com/us-en/shop/tech-takes/what-are-computer-cookies
https://www.lawinsider.com/dictionary/uidh
https://www.linkedin.com/pulse/how-does-verizons-supercookie-work-alexander-k-senemar
https://www.seobility.net/en/wiki/Session_ID

WHY DNS?

The Domain Name System (DNS) brings human-readable text to the infrastructure of the Internet by allowing regular language constructs to be used for navigating the complexity of the underlying computer network of networks. You can use regular language instead of large sequences of digits.

It largely evolved from a file called HOSTS (which still appears on most OSes) used in the pre-Internet network called ARPANET. The local HOSTS file was used by every computer for addressing other network devices, but the ARPANET network was so small that no naming resolution was required at that time. It was also designed in this fashion (in addition to other factors) so the network as a whole could theoretically survive communication outages caused by thermonuclear war through rerouting data, and going around any individual set of nodes that might have been obliterated.

DNS is built on top of a system of cryptic multiple-digit code formats often referred to by their format as IPv4 (or the newer IPv6), each representing digital computer network addresses based on Ethernet and Internet Protocols. Basically, the intent of DNS is to make the connections between the world's network of computer networks readily navigable by any human at any keyboard.

The DNS global directory starts with top-level domains (TLDs) with regular language alphabet and words overlays divided up into named categories such as .com, .gov, and .edu. This TLD design extended to other nations, allowing local language use for domains by adding the country code after the English TLD, viz., example.com.eu.

To do this, DNS maintains entries for each record with a specific look-up for the values in the record. These look-ups allow for standardized, human-sensible phrasing (e.g., firstexample.com, anotherexample.com, lastexample.com) onto that underlying IP infrastructure and configuration, in the following formats:

IPv4 and IPv6 Networks

Via IPv4

The IPv4 address is constructed using base 10 or decimal numbers (here represented with an "x") in each character position of the form:

xxx.xxx.xxx.xxx

For IPv4 example: **google.com** maps (as of Q2 of 2024) via DNS A record (i.e., IPv4) to 207.126.144.0.

There are $2^{32} = 4{,}294{,}967{,}296$ IPv4 addresses available in total.

Via IPv6

The IPv6 address is constructed using base 16 or hexadecimal numbers (here represented with an "n") in each character position of the form:

nnnn.nnnn.nnnn.nnnn.nnnn.nnnn.nnnn.nnnn

For IPv6 example: **google.com** maps (as of Q2 of 2024) via DNS AAAA record (i.e., IPv6) to 2001:4860:4000:0:0:0:0:0.

There are $2^{128} = 340,282,366,920,938,463,463,374,607,431,768,211,456$ IPv6 addresses available in total.

Those number schemas represent the Internet protocol address of Google's servers (given Google's scale, it actually has many, many such DNS records containing many such addresses and servers).

You can use this difference between IP and domain names for troubleshooting when a server might be down, when your DNS might be down, or when you lack an Internet point-of-presence (PoP). For example, ping a server from a terminal window using either the IP number directly or the domain name (in this latter approach, the request is routed through the DNS entry for the site).

A DNS entry may contain dozens of various records but the basic entries are around IP location and information details (see below):

Example entries in a DNS record

Record Name	Record Description
A	The record that holds the IPv4 address for a domain.
AAAA	The record that contains the IPv6 address for a domain.
MX	The name of the email server for the domain.
TXT	Any special comments about the domain.

Due to its massively distributed nature and extensive use of caching across the system, the DNS network sometimes has issues around specific record updates as they propagate across the networks of networks around the globe. This is where issues like propagation of entries (for use in other locations), duplication, and conflicts rapidly become complex topics. In general, a change in or creation of a DNS entry must propagate globally to become fully in effect or available.

A common analogy has DNS as similar to a telephone directory system, where everyone has a phone book and instead of keying in numbers, you type in the name of the computer (server) you want to get to. However, this analogy is misleading: DNS has no single global authority, and is generally provided as a local copy by the ISP providing your Internet PoP. It is not run by a for-profit corporation. You can, in fact, configure and use your own private and local DNS, if you want to.

DNS is a massively distributed system, so that no central world authority adjudicates every directory entry (although generally, there are smaller local authorities for every individual zone serviced).

FURTHER READINGS

https://blog.hubspot.com/website/what-is-dns-server
https://datatracker.ietf.org/doc/html/rfc1035
https://robots.net/tech/how-to-connect-to-ip-address/
https://www.arin.net/resources/guide/ipv6/
https://www.cloudflare.com/learning/dns/dns-records/
https://www.cloudflare.com/learning/dns/dns-records/dns-a-record/
https://www.cloudflare.com/learning/dns/dns-server-types/
https://www.digitalocean.com/community/tutorials/an-introduction-to-dns-terminology-components-and-concepts
https://www.lifewire.com/what-is-the-ip-address-of-google-818153
https://www.top10vpn.com/tools/what-is-my-dns-server/

WHY ARE THERE IPV4 AND IPV6?

IPv4 originated from a series of communication protocols used in ARPANET, a Pentagon research project. Starting in 1969, the development of the Internet Protocol (IP) and Transmission Control Protocol (TCP). IP was used as an address system for network connections, and TCP for packet ordering and error checking. It wasn't until 1977 that TCP/IP was split into two distinct bodies, and it took three more years until the first public edition of IP—the fourth version of IP, hence IPv4—was created in 1980.

IPv4 was created to allow packet exchange on computer networks, enabling computers to communicate with each other and exchange data. IPv4 has a 32-bit address format consisting of four octets—four decimal brackets of eight bits each—where each octet is represented by a number from 0 to 255. Refer to the answer on DNS for more information about formatting and usage. This addressing system allowed a single network to uniquely identify around 4.3 billion devices connected to it, which soon became a concerning limit by the late 1980s and especially the early 1990s. Network Address Transition (NAT) technology aids in extending the lifespan of IPv4 addresses by mapping several internal devices to a single externally facing IP address. By doing so, it counteracts the sheer volume of IP addresses that would otherwise be required.

IPv6 began development in the mid-1990s. By 1999, Internet registry and number assignment authorities began assigning and allocating IPv6 address blocks. IPv6 uses 128 bits—approximately 2^{128} combinations of digits, or 340 trillion trillion trillion addresses—so its sheer size effectively remediated IPv4's address allocation limit. Contrary to what one might think, however, IPv6 will not entirely replace IPv4. According to a Google survey from August 2024, only about 45% of its global user base has IPv6 connections available. The two address systems will coexist for the foreseeable future due to various limitations with device hardware, financial logistics, and IPv6-incompatible devices.

Decimal	Keyword	Protocol
0	HOPOPT	IPv6 Hop-by-Hop Option
1	ICMP	Internet Control Message
2	IGMP	Internet Group Management
3	GGP	Gateway-to-Gateway
4	IPv4	IPv4 encapsulation
5	ST	Stream
6	TCP	Transmission Control

Currently, IPv4 is generally simpler and cheaper to maintain. Many enterprise companies are comfortable maintaining their older, dependable machines on

IPv4, especially because the transition to IPv6 isn't a pressing issue. Thus, they avoid the logistical headache of upgrading countless devices, hiring or training new people to work the technology, and handling all the possible connectivity problems with IPv6. Not to mention, NAT64, a translation technology for IPv4 and IPv6, will vastly extend the lifespan of NAT and mitigate the urgency for upgrading. It could be years, or even decades, before a total transition to IPv6 would be entirely feasible. According to Jared Mauch, an internet expert at an ARIN (American Registry of Internet Numbers) panel on IPv6: "when you talk about product lifecycles, probably about 25 years."

Historical aside: the intermediate IPv5 was never truly an internet protocol. Instead, it was commonly termed ST (for "stream"), then ST2, a streaming protocol designed to be encapsulated in regular IPv4 packets as real-time packets. These packets would be for communications and media transmission, like voice and video. Their protocol number, an identifying number for internet protocols, was 5—so even though ST2/IPv5 was never actually regarded as "IPv5," the developers of IPv6 still decided to skip over the number 5 to avoid any protocol number conflation.

FURTHER READINGS

https://1nce.com/en-us/resources/iot-knowledge-base/iot-connectivity/cellular-networks/what-is-nat
https://bluecatnetworks.com/blog/ipv4-vs-ipv6-whats-the-difference/
https://ipv4marketgroup.com/a-brief-history-of-ipv4/
https://www.arin.net/blog/2022/05/03/arin-49-ipv6-panel/
https://www.arin.net/blog/2022/05/12/other-ip-version-numbers/
https://www.cisco.com/c/en/us/support/docs/ip/network-address-translation-nat/217208-understanding-nat64-and-its-configuratio.html
https://www.cloudflare.com/learning/ddos/glossary/tcp-ip/
https://www.google.com/intl/en/ipv6/statistics.html#tab=per-country-ipv6-adoption

WHY IS LATENCY STILL SUCH A HUGE PROBLEM?

Latency in networking is how long it takes data to travel from a network device source A to a network device destination Z across a network or networks of devices. Latency is a real-time delay, which for computer networks is usually measured in milliseconds, between an action or input and a response or effect. Latency is everywhere, even if it is so apparently minimal, it is necessary to measure it in increments like femtoseconds. It is important to calculate because even the smallest of latencies can compound and build up into huge delays or stalled operations.

Sometimes it is possible to obscure the UX of annoying latency by hiding a network latency inside another one that does not impact the UX. As an example, Instagram had to factor latency into some of its early software design decisions: Instagram's founders learned that users were annoyed by the time it took (wireless 2G networks back then) to upload photos. The founders decided to change product function to take the photos the user had already tagged and upload to their servers while the user was typing their Instagram message, thus giving a UX to the user of immediate photo availability. Availability of the photos, already loaded, became a much-loved feature of Instagram.

Common UXs of latency on a personal computer include

- clicking on a webpage button and waiting until the action occurs;
- playing a video game and your RPG character firing a weapon at an enemy player, which doesn't get hit, because bad latency or lag causes a delay between your firing the shot and the server calculating its destination;
- delays on video and audio synchrony, real-time stock tickers, movie playback, file downloads.

Latency exists between and within every connection of every device, circuit, peripheral, and network. Currently, the optimum latency speed known to humans to exist between two connected objects is determined by the speed of light (i.e., the c in the famous $E=mc^2$ equation and currently about 300,000 kilometers per second or about 186,000 miles per second). No connection can have a better latency than that provided by the speed of light, according to the prevailing understanding of physics. The speed of light thus governs all communication connections of any distance (we are excluding the discussion of physics hypotheses such as tachyon particles, which are posited to travel faster than the speed of light, or the impact of instantaneous communications via any applicability of Bell's Theorem).

Aside: It is important to note that the speed of light is generally considered to be universal, so this same latency exists in your personal experience of the world. Your vision is also constrained by this same speed-of-light latency. For example, it is generally known that the stars seen in the sky are as they existed hundreds of millions of years ago. That's latency from travel time, even at the speed of light. However, even the view you see out the window is actually an image from a fraction of a moment ago rather than the reality that exists out there now. So, you are technically always seeing the past.

Latency is everywhere. It is persistent particularly in computer networking (see below for important terms used for discussing latency in networks).

Important network terms used for discussing latency:

- PING—a number, usually measured in milliseconds, that reflects the amount of time required to get from point A to point B across a network. Ping is a number in aggregate that tells you the current overall communication time between your position and the location of your destination (usually a server of some sort). Ping is additive. It is composed of all of the pings involved between the network device connections, making up the network stops between point A and point B. In the case of gaming, it is the elapsed communication time between your gaming client and the gaming server. Ping can vary by second, by hour, by day, by event, by outages, and other environmental factors.

- HOPS—the number of stops or device traversals through a network of devices that a connection goes through to get from starting point A to destination point Z. A hop represents a single connection between two network devices on that journey.
- LAG—any delay in computer communication.

The command traceroute is often used in illustrating latency in a computer network. It tracks the hops between a source network device (in the case of the Traceroute example below, the source device is one of the authors' computers) and a destination network device (in this case, an MIT computer server).

The traceroute command relies significantly on the use of time-to-live (TTL), which is not detailed in this answer (it is a textbook network setting and readily found online). A traceroute duration is largely determined by the aggregate time of connecting the hops (connections between intervening computers).

Traceroute example (provided by authors) current as of 2025

```
% traceroute allspice.lcs.mit.edu
traceroute to mercury.lcs.mit.edu (18.26.0.122), 64 hops
max, 40 byte packets
1 192.168.4.1 (192.168.4.1) 4.114 ms 3.238 ms 4.312 ms
2 10.0.0.1 (10.0.0.1) 4.087 ms 3.901 ms 3.465 ms
3 100.93.15.67 (100.93.15.67) 13.422 ms
        100.93.15.66 (100.93.15.66) 14.951 ms
        100.93.15.67 (100.93.15.67) 17.089 ms
4 po-312-405-rur101.exeter.nh.boston.comcast.net
  (68.86.236.37) 15.523 ms
        po-312-406-rur102.exeter.nh.boston.comcast.net
  (96.108.101.89) 18.450 ms
        po-312-405-rur101.exeter.nh.boston.comcast.net
  (68.86.236.37) 14.906 ms
5 po-100-xar02.exeter.nh.boston.comcast.net
  (162.151.151.101) 14.772 ms
        po-100-xar01.exeter.nh.boston.comcast.net
  (162.151.151.225) 13.682 ms
        po-100-xar02.exeter.nh.boston.comcast.net
  (162.151.151.101) 16.287 ms
6 * be-301-arsc1.woburn.ma.boston.comcast.net
  (162.151.151.65) 22.858 ms *
7 be-501-ar01.woburn.ma.boston.comcast.net (162.151.52.78)
  18.389 ms
```

```
      be-501-ar01.needham.ma.boston.comcast.net
  (162.151.52.34) 16.527 ms
      be-502-ar01.woburn.ma.boston.comcast.net
  (162.151.52.90) 17.170 ms
8 ae-0-eg-bstpmall74w.boston.ma.boston.comcast.net
  (68.86.238.34) 39.689 ms
      ae-1-eg-bstpmall74w.boston.ma.boston.comcast.net
  (162.151.113.10) 15.023 ms
      ae-0-eg-bstpmall74w.boston.ma.boston.comcast.net
  (68.86.238.34) 18.354 ms
9 50.201.57.174 (50.201.57.174) 17.072 ms 17.807 ms
  15.399 ms
10 dmz-rtr-1-external-rtr-3.mit.edu (18.0.161.13) 19.526 ms
   15.985 ms 17.374 ms
11 dmz-rtr-2-dmz-rtr-1-1.mit.edu (18.0.161.6) 18.217 ms
       dmz-rtr-2-dmz-rtr-1-2.mit.edu (18.0.162.6) 17.804 ms
       dmz-rtr-2-dmz-rtr-1-1.mit.edu (18.0.161.6) 17.288 ms
12 mitnet.core-1-ext.csail.mit.edu (18.0.162.142) 19.742 ms
   14.752 ms 19.057 ms
13 * * *
14 trantor-int.core-1.csail.mit.edu (128.30.4.2) 21.364 ms
   23.691 ms 19.956 ms
15 mercury.lcs.mit.edu (18.26.0.122) 17.158 ms !Z 18.842 ms
   !Z 15.977 ms !Z
```

It is possible for a single slow ping at one hop to cause the overall connection to slow or even become temporarily unavailable. Without analysis on a per-hop basis, the overall ping will be perceived by the user to just be very large or, even, untenable for use.

Generally, a failed hop will eventually result in redirection around the failed network device. While said redirection is built into the survive-atomic-war-design of the original internet, it will not always occur in short timeframes (e.g., a few hours) and thus may not seem to occur to a user as needed in the moment. It is also possible to have the aggregate of hops and pings make for a very large total ping between start and end points, particularly if there are many hops involved. This can also be seen in connections to some countries with initial points of presence requiring a large number of hops (generally any number in excess of 64 hops) or additionally slow entry points—mainly seen with less developed countries—which will result in large latency times.

In 2010, a Google employee (named Jeff Dean) presented at a conference a handy reference piece around how latency and timescales and their impact on computing device behaviors—and basically what every developer should know and understand about latency. This "rule of thumb" list remains pertinent to programmers and Computer Scientists today.

For example, when looking at a system implementation, a developer should for any system at scale keep in mind practical transit details such as "6–7 round trips between Europe and the US per second are possible…" These sorts of constraints will impact product and system design and behavior, especially when systems are implemented at scale. The data in a version of that document was updated in 2019 here: https://www.freecodecamp.org/news/must-know-numbers-for-every-computer-engineer/

The reader should be advised that the times listed will get increasingly shorter but will arrive at a limit as they approach the value for the latency inherent in the speed of light.

FURTHER READINGS

https://investlikethebest.libsyn.com/kevin-systrom-and-mike-krieger-how-to-build-a-great-product
https://static.googleusercontent.com/media/sre.google/en//static/pdf/rule-of-thumb-latency-numbers-letter.pdf
https://www.cloudflare.com/learning/performance/glossary/what-is-latency/
https://www.reviews.org/internet-service/what-is-internet-latency/

WHY DO COMPUTING DEVICES HAVE MAC IDENTIFIERS?

A Media Access Control (MAC) identifier allows device routing identification to support many functions, most commonly video streaming, for a device regardless of the network's varying or underlying IP assignments for that device.

The MAC identifier, or address, is stored within the networking hardware on any computers, Wi-Fi adapters, and embedded devices that have networking capabilities and, unlike global IP addresses, is not assigned by any external governing body like ISPs (Internet Service Providers. For more info on what those are, see our answer for "Why DNS?"). The id is used to supply a unique network identifier for interacting specifically with the device (as the device may or may not always be consistently at the same IP address it was originally associated with). There are 2^{48} possible MAC addresses, which are usually universally assigned by the device manufacturer.

MAC ids are a pre-existing assignment of a MAC id and broadcasts or self-identifies the id at connection time. A device may have multiple MAC ids over the course of its lifetime.

The MAC id is usually constructed using base 16 or hexadecimal numbers of the form:

nn:nn:nn:nn:nn:nn

There are subtypes and administrations of MAC ids also, for wireless routers and other devices. These are sometimes referenced by SSID and BSSID. See the answer to "Why are there so many different ways to identify devices on a network (MAC, IP, SSID, etc)?"

WHY ARE THERE SO MANY DIFFERENT WAYS TO IDENTIFY DEVICES ON A NETWORK (MAC, IP, SSID, ETC.)?

All of the different layers and ways of identifying a device are to solve various problems that crop up once you start building networks of devices. You have the network's physical layer, the logical layer, etc.

Each addressing method handles different problems associated with having devices on a common network such as the device's network location, device's logical use, routing, and so on. Many of the most commonly used device identifiers and their background are listed below.

Ethernet addresses are designed to access different devices on the same physical network. It gets the name from Bob Metcalf, the inventor of Ethernet, who used a 19th-century physics concept to show that his network system for connecting early devices could be applied to network any computing device.

The generally accepted story about the creation of the Ethernet name is: Metcalf had designed a networking system to connect a brand of microcomputer called Altos Computers, so it was called Altos-net.

But "Altos-net" did not capture the full possibilities Metcalf had envisioned. The network could connect more than Altos devices. He wanted a name that reflected that broader applicability and use.

In physics, in the latter half of the 1800s, there was a concept used to reconcile the lack of a physical medium in the propagation of electromagnetic waves from one physical object to another. The problem was that this propagation appeared to occur through empty space, which was not acceptable to physics theories. So, physicists at the time postulated the existence of an invisible yet all-present medium called the "luminiferous ether" existing between all physical objects.

Metcalf borrowed that "ether" to indicate his network design worked as a connecting medium for any device.

IP is used to handle routing and provide an abstraction layer over the physical (Ethernet) network. There are two current field versions: IPv4 and the more recent IPv6. The difference between them is the staggering number of devices or points they can enumerate.

IPv4 and IPv6 Networks

Via IPv4

The IPv4 address is constructed using base 10 or decimal numbers (here represented with an "x") in each character position of the form:

xxx.xxx.xxx.xxx

For IPv4 example: **google.com** maps (as of Q2 of 2024) via DNS A record (i.e., IPv4) to 207.126.144.0.

There are $2^{32} = 4{,}294{,}967{,}296$ IPv4 addresses available in total.

Via IPv6

The IPv6 address is constructed using base 16 or hexadecimal numbers (here represented with an "n") in each character position of the form:

nnnn.nnnn.nnnn.nnnn.nnnn.nnnn.nnnn.nnnn

For IPv6 example: **google.com** maps (as of Q2 of 2024) via DNS AAAA record (i.e., IPv6) to 2001:4860:4000:0:0:0:0:0.

There are $2^{128} = 340{,}282{,}366{,}920{,}938{,}463{,}463{,}374{,}607{,}431{,}768{,}211{,}456$ IPv6 addresses available in total.

MAC id

is designed to allow for correct device identification for use across wireless networks (for example, in video streaming, where one device is switched for another by the user). The MAC id is constructed using base 16 or hexadecimal numbers of the form:

NN.NN.NN.NN.NN.NN

Every device you have, depending on age, is likely to have a MAC id.

IMEI uniquely identifies a mobile device. Generally, the IMEI is of the form:

XX-XXXXX-XXXXX-X

although an IMEI is not *necessarily always* in this format, due to compatibility across versions occurring over time, and implemented in evolving devices and technologies and the introduction of newer versions. Unlike other identifiers, it can be used to blacklist a specific device (for example, for a stolen phone), among other uses.

Your mobile device and occasionally your digital watch will have IMEI codes but generally your laptop and tablet will not.

SSID—Service Set Identification Device—a named device per the IEEE 802.11 wireless networking, where the said device provides a known bundle of network services.

BSSID—Basic Service Set Identifier provides identification of the wireless access point in a Wi-Fi network derived from the 48-bit MAC address system.

ICCID—The identifier (generally 18—24 digits in length) for the SIM card, which holds the phone number for the mobile device.

What have we been solving for here? You want to connect a device to a network. So the reasoning here is that if you want to identify a device on an Ethernet network, you reconcile to the IP. The IP on Ethernet is intended to give you a unique way of addressing the device. But what if it's on its own subnetwork and the network itself has the public IP? Then you need a way to identify the device within the network so you can tag the device specifically. This is all straightforward while the network is wired. What if it's wireless, like a Wi-Fi network? And your device needs to connect to the network but also be addressed on the network. Or if it's wireless but in the form of a mobile network?

Each one of these situations was at first an edge case that needed to be resolved to manage a specific device situation. However, radical growth in the use of computing devices—fixed, ambulatory, laptop, tablet, phone, etc.— spawns a primary use case from what was originally an edge case as the type of edge case grows in maturity.

As happens in computer science, what starts as an edge case can become primary use cases over time as technologies improve or mature. Hence, the proliferation of what appear to be similar identifications over time becomes dominant in the use case for which they were designed. In reality, these identifiers are not the same at all because each identifier has a very specific applicability and unique use case to solve.

Acronym Table of various device identifiers:

Name or Acronym	Referent
BSSID	Basic Service Set Identification Device
Ethernet	"Luminiferous Ether" Net
IMEI	International Mobile Equipment Identifier
ICCID	Integrated Circuit Card Identifier
IP	Internet Protocol (IPv4, IPv6)
MAC id	Media Access Control Identifier
SSID	Service Set Identifier Device

FURTHER READINGS

http://www.ethermanage.com/why-is-it-called-ethernet/
https://en.wikipedia.org/wiki/International_Mobile_Equipment_Identity
https://www.freecodecamp.org/news/the-complete-guide-to-the-ethernet-protocol/
https://www.geeksforgeeks.org/what-is-internet-protocol-ip/
https://www.imei.info/faq-how-check-IMEI/
https://www.imei.info/faq-what-is-IMEI/

WHY IS PUBLIC KEY AND PRIVATE KEY ENCRYPTION SO POPULAR IN ONLINE COMMUNICATIONS?

PKI is extremely popular in securing online communications for a number of reasons. The shortest answer is that not only is it incredibly difficult to break the encryption, but it also prevents decryption via "Man in the Middle" (MitM) attacks. In other words, an attacker with the full ability to observe the connection between two computers using PKI would not be able to decode their messages, even if that attacker witnessed the very first establishment of the PKI info between them!

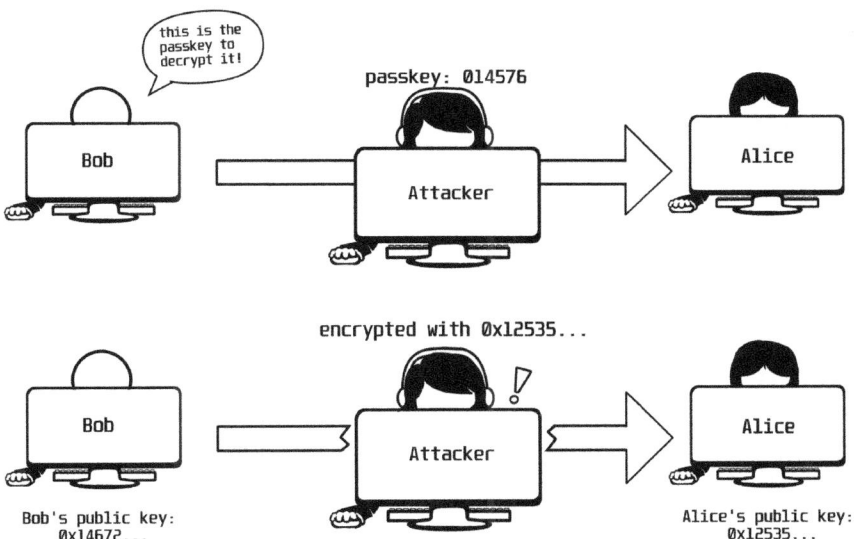

If you've never explored this form of cryptography before, you might think of encryption as involving some kind of transformation routine, and a special passphrase or unique number to apply with the routine that makes the encrypted result especially unique. For instance, a basic encryption routine might be to take the word "hello," convert it to its ASCII equivalent numbers, and then add a special secret value "04581" to each of them.

The issue, then (besides the fact that this is very easy to break), is: how does the sender convey the special value to their recipient without giving it away to an attacker listening in on their communications? When we connect to secure pages or log in to our inboxes via encrypted communications for the first time, we don't separately mail encryption keys back and forth to the companies hosting those websites.

PKI encryption is a completely different approach where no secret value or key has to be transmitted between communicators. Each person has their own two special values, a public key and a private key. The public key is available to everyone and is used for encryption. The private key is used for decryption and is never shared with anyone else.

If Bob wants to send a message to Alice, he fetches Alice's public key, encrypts the message, and sends it along. Alice uses her private key to decrypt it. Although the public key was used to encrypt the message, it can't be used to break the encryption! Only the private key can do this—the public key doesn't reveal any information about the encryption. Therefore, there's no password swap, no secret value exchange—any two people can encrypt and transmit messages to each other, under full observation by an attacker, without any worry of the attacker breaking that system.

So you may wonder... how is this even possible? How can two numbers, clearly related to each other via encryption and decryption, not be used to reveal each other?

The full explanation for this requires a fair amount of number theory, which is outside the scope of this answer, so we will just present a few underlying concepts. If you want to delve more into the math behind PKI encryption, we recommend the following textbook, which provides a great overview from beginning concepts all the way up to implementing PKI encryption: *Information Security: Principles and Practice* by Mark Stamp, Chapter 4.

Alternatively, you can search for Kid-RSA, a simplified version of the algorithm that demonstrates the concepts. We don't have a particular book or paper recommendation to get into this one, as it is a popular component of match courses and open-source coding repos, and numerous sites and PDFs are available for it online.

Here, we present a few underlying mathematical concepts behind PKI encryption—one-way functions and modular arithmetic.

One-way functions are operations that, given an input, produce an output that is very difficult to reverse. In PKI encryption, these functions typically incorporate modular arithmetic to protect from reversal.

Taking the modulo, or mod, of a number means dividing by it and taking the remainder. For instance, 5 mod 3 is 2 (5 divided by 3 has a remainder of 2). 8 mod 7 is 1. Modding by a number that divides evenly is 0, as there is no remainder—for instance, 22 mod 11 is 0.

If you have some hidden number X mod 7 = 5, it's very difficult to determine what X is, because there are infinitely many possibilities. X could be 12, 19, 26..., and so on. If you were to encode a value using a modulo, an attacker would just have to brute force until they uncovered the original value.

Another one-way operation is, given a gigantic number, determining quickly which two numbers multiplied together form that number. Picking a large number (with say, 15+) digits can yield hundreds to thousands of possibilities, and it's not clear which two numbers are correct, other than by guessing. Choosing prime numbers as your product inputs makes the problem even harder, as this means the product can't be reduced to a smaller factor before brute forcing.

PKI encryption uses both of these methods to simultaneously generate two special values, a public key and a private key. The public key is based on multiplying together two large prime numbers, which must be kept secret. The private key is derived from information relating to these prime numbers, such that applying the private key to an encrypted message decrypts it. Because this entire derivation process is based on these one-way functions, it's nearly impossible to reverse it and derive the private key used to decrypt messages. Similarly, the public key itself cannot be used to decrypt anything.

HISTORICAL ASIDE

One of the most famous and earliest examples of PKI which is still used today is RSA, short for Rivest-Shavir-Adleman, after its founders. Adi Shamir, Ron Rivest, and Leonard Adleman are three MIT researchers who developed RSA to solve exactly the problem we introduced this entry on: safe transmittal of information without a key exchange in the 1970s. Amazingly, just a few years before them, an English mathematician named Clifford Cocks, working for the British intelligence agency, derived the same process, but the implementation of it was shelved for being too expensive to adopt on the limited hardware available to computers at the time. It wasn't until the project was declassified in 1997 that the project was revealed as a nearly simultaneous discovery of public key encryption.

WHY DO TLS/SSL CERTIFICATES EXPIRE?

TLS certificates exist for website encryption and authentication. When a TLS certificate is assigned, it becomes a snapshot of the regulations at the time, as well as the website information it authenticated. This information normally contains data like the domain and owner names, the certificate authorities (CAs) that assigned it, and its date of expiry. Therefore, it's important for certificates to be kept current to reflect any updates to website information, owners, or security regulations.

An indefinite TLS certificate lifespan would pose several security risks. For example, imagine if a third party hijacked a company's TLS certificate and domain name. It could pass itself off as the legitimate site, jeopardizing the security of anyone who visited or entered sensitive information. By keeping certificate life spans shorter, it's simpler for both website users and owners to confirm the currency of their site's security.

Companies don't always pay good attention to their certificates, however. On December 6, 2018, Telefonica, one of the largest providers of mobile services in the world, saw its 4G services go down everywhere. The outage affected upward of 32 million people globally, not just direct customers of Telefonica's services, but also any users whose activities passed through Telefonica's network channels. It went down for the entire day. The network malfunction was traced back to Ericsson, a giant in infrastructure and network management systems, and a single expired certificate. A single TLS certificate expiration caused a global network service meltdown.

The length of certificate lifespans is a growing security concern. In 2020, Apple, Google, and Mozilla announced that TLS certificates issued after September 1 could not exceed a life span of 398 days. In March of 2023, Google shared its intentions to move to a 90-day certificate policy. In theory, this would encourage transition from manual to automated certificate renewal, essentially making certificate expiration obsolete. However, this shift would involve a total rework of how certificates are handled. It wouldn't just change from manual to automatic, but also from relatively lax certificate management to a rigorously enforced set of procedures.

FURTHER READINGS

https://blog.mozilla.org/security/2020/07/09/reducing-tls-certificate-lifespans-to-398-days/

https://cdn.preterhuman.net/texts/computing/security/SSL%20And%20TLS%20Essentials%20-%20Securing%20The%20Web%202000.pdf

https://https.cio.gov/technical-guidelines/

https://support.apple.com/en-us/102028

https://www.chromium.org/Home/chromium-security/root-ca-policy/moving-forward-together/#:~:text=Chromium%20%3E%20Chromium%20Security%20%3E%20Root%20Program%20Policy,However%2C%20there%E2%80%99s%20still%20more%20work%20to%20be%20done.
https://www.cloudflare.com/learning/ssl/transport-layer-security-tls/
https://www.cloudflare.com/learning/ssl/what-happens-in-a-tls-handshake/
https://www.cloudflare.com/learning/ssl/what-is-ssl/
https://www.crowdstrike.com/blog/the-risks-of-expired-ssl-certificates/
https://www.forbes.com/sites/daveywinder/2018/12/07/here-is-the-ridiculous-reason-32-million-telefonica-o2-users-waved-goodbye-to-4g-data-yesterday/
https://www.sectigo.com/resource-library/why-certificates-expire-shorter-validity-periods
https://www.ssl.com/guide/tls-standards-compliance/

Programming

This section takes an alternative look at popular programming language features, explaining the reasons behind their particularities, how certain language features are implemented, and the motivations behind using one type of language or another, in a way that is helpful and practical to the average software development student.

DOI: 10.1201/9781003519379-5

WHY DO MANY LANGUAGE ARRAYS START AT 0?

There are historical and often quite technical debates over arrays starting at 0. Some languages like Mathematica, Lua, Fortran, and Julia use 1-based indexing. At this point in time, a far larger number of the many many languages now in existence use 0 as the starting index of an array, but the debates are still active.

We won't get into those debates nor examine the technical definitions of indexes and counts, which can be found in many textbooks and online. Nor will we review the potential inefficiencies of the 1-based array for compiling or use in assembly. Those discussions can also be found in textbooks or online.

The heart of the usual arguments against 0-based arrays is the assertion that folks did not really start counting at 0 prior to computer programming. But this is quite untrue. Examples are provided below.

The distinction in the argument between 0 and 1 that we bump up against here is the mathematical identity of cardinal numbers (which are counted, as in 1, 2, 3,...) and ordinal numbers (which are ordered, as in 1st, 2nd, 3rd,...). Alternatively, you can refer to these different types colloquially as numbers and numerals. (Using the classic computing example from the fencepost problem as a reference, you could say the post is a number and the fencing is a numeral.)

This distinction impacts any type of thing that can be measured by offset or duration. The examples of those things are, in fact, fundamental to how we navigate and document the world. And, contrary to the 1-based argument about no historical use of 0-based indexing, we've been using 0-based indexing since the beginning of time!

Time

The example of time is a good place to start. The beginning of the day starts at midnight or 00:00. While that first midnight hour can be dark indeed, you haven't reached hour 1 until you're into the *second* hour of the morning. That first space of an hour is a 0, and the second space of an hour is a 1. For the 1-based argument, the confusing impact of trying to start midnight at 1 and proceed from there is left as an exercise for the reader.

Distance

Take distance. Using kilometers, imagine a place and make it ten kilometers in size. As we traverse this distance, this array of space, we reach the first kilometer (i.e., a number) after the space of that first kilometer has passed (i.e., a numeral). Where are you in the first kilometer of a journey? That first kilometer has a cardinal value of 0 (a number) and an ordinal value of first (a numeral).

Age

Or let's define an array called Childhood and have it contain successive years, 12 in total, and let's begin with the first year when you are not even 1 year old. How old are you in your first year of life? Most parents will avoid any problem by citing elapsed weeks or months since birth.

So, we use the array's index as a type of offset into the array. This is different from using a number identifying, say, a slot in a cubbyhole (for example, the slot for a specific apartment number in an apartment mailbox cluster). As every C programmer has learned, in arrays, 0 is first while the number 1 is second, and so on.

Yet, those who argue for index-0 have a very strong basis in practice: the fact remains that human beings in general start counting at one. This may be likely as we don't recognize zero fingers! So, the debate about indices and arrays and where to start can be counted on to remain vibrant.

FURTHER READINGS

https://cseducators.stackexchange.com/questions/5023/why-do-we-count-starting-from-zero

https://medium.com/@mumaticha/array-indexing-why-are-arrays-indexed-from-zero-8a367c04e3e

https://stackoverflow.com/questions/18109671/why-doesnt-the-index-list-of-an-array-begin-with-1?noredirect=1&lq=1

https://www.cs.utexas.edu/~EWD/transcriptions/EWD08xx/EWD831.html

HOW DO COMPILERS KNOW WHERE TO FIND IMPORTED FILES AND LIBRARIES?

Have you ever wondered, when you import a library like stdlib into one of your C++ programs, how the compiler "knows" where to find it? How does it find these files, versus those you write yourself and then include?

Operating systems have default locations for commonly shared files like libraries. When you import a library, your compiler or interpreter sweeps these locations for any files that match the specified name. As for importing user-written local files in either the same directory as the program or a custom location, typically the compiler or interpreter will sweep the program's directory and, optionally, any locations specified by the user on the command line. A standard set of include paths for Linux includes /usr/local/lib, /usr/local/lib64, and /usr/lib.

Compilers have support for displaying the current list of search paths for libraries and header files. For instance, with the Clang compiler, you can add the flag "-v" to a typical compilation command to print a list of the include paths it will search. Below is some sample output from printing the search paths on MacOS:

Clang Search Paths—MacOS

```
/usr/local/include
/Library/Developer/CommandLineTools/SDKs/MacOSX.sdk/usr/include/c++/v1
/Library/Developer/CommandLineTools/usr/lib/clang/15.0.0/include
/Library/Developer/CommandLineTools/SDKs/MacOSX.sdk/usr/include
/Library/Developer/CommandLineTools/usr/include
/Library/Developer/CommandLineTools/SDKs/MacOSX.sdk/System/Library/
  Frameworks ( framework directory)
```

It's handy to have an awareness of these locations, as sometimes you might have errors importing a library and need to double-check that it has been installed correctly. If you ever get odd dependency or type errors stemming from your import statements, you might need to peer through some of the corresponding included files to troubleshoot them. And if at times you're trying to install some packages and the installer just isn't quite working right, you may have to download some source files for that package and place them in the correct locations yourself. Knowing the default paths, or at least knowing enough to look them up, can be a tremendous boost in these situations.

WHY HAVE DIFFERENT PARADIGMS IN PROGRAMMING LANGUAGES, LIKE IMPERATIVE VERSUS FUNCTIONAL?

Programming paradigms are the fundamental design decisions behind programming languages. Object-oriented and imperative designs are a couple that the average reader may be most familiar with: imperative style consists of a set series of actions which modify state, with statements like "$x=5; x=x+7$." Object oriented is a form of imperative programming that, among other things, involves representing objects, their attributes, and methods. Popular languages with both paradigms include C++, Python, and Java.

Oftentimes, programming students and junior developers will develop a decent familiarity with these paradigms, but wonder what major useful differences exist between this and others used in industry settings: in particular, functional languages, which tend to be one of the other most popularly used. To make things even murkier, most languages include support for functional, imperative, and object-oriented styles of programming all together. There's much detail we leave out in an effort to convey some of the core ideas of functional programming without overwhelming the reader.

As we mentioned earlier, object-oriented paradigms allow for creating objects that can represent different things. These objects can have any amount of customizable data, as well as state—statuses, counters, and so forth. An important trait here is that this data can be *mutable*, or in other words, it can be changed over the course of program execution. For instance, let's say you have a small business selling puzzles, and you write a program that keeps track of orders for puzzles you sell. You could define a Customer class that has attributes that describe the state of the transactions between you and someone ordering items, like outstanding balances, as well as functions that correspond to various actions like purchases and refunds, which update these attributes. The example below shows some code for this.

Pseudocode for a class to keep track of shop orders:

```
class Customer {
        String name;
        int outstandingBalance = 0;
        int credit = 0;
        Puzzle[] shoppingCart = [];
        void purchaseItem(Product selectedItem) {
                self.itemsToSend.append(selectedItem);
                self.outStandingBalance += selectedItem.cost;
        }
        void purchaseGiftCard(int giftCardAmount) {
                self.credit += giftCardAmount;
        }
}
```

```
        void processPayment(int paymentAmount) {
            self.outStandingBalance -= paymentAmount;
}
```

The Customer class has two methods to update some information about it when purchases occur. When a particular action takes place, the state of the object is updated; for instance, when a customer makes a purchase, their outstanding balance is increased. When the customer makes a payment, the balance is updated similarly. These functions have something called *side effects*, **which is that they change some data located outside of the function when they're called**. In this case, this is intentional and helpful for the purpose of the program. There are some pitfalls to this as well. Let's say we return to this program later and want to change the program to support purchasing gift cards and keeping track of a store credit. If we add a credit attribute, this is another piece of state we have to maintain and modify while keeping in mind how it might affect the outstanding balance, and we'll likely have to update all functions that work with this attribute to take this credit into account. Any time we want to add new functions that affect the balance or credit on that account, we have to review the other functions and make sure we aren't introducing any conflicts or running into any unexpected side effects. This is a simple example, but in more complex programs with many types of objects interacting with each other, developers may have to spend a significant amount of time familiarizing themselves with the software, classes, and the flow of changes in states before making any changes.

Functional programs take an entirely different approach: in these, actions and changes in data are the result of functions (hence the name). In a completely functional design, all data is immutable. Instead of modifying data, functional programming involves creating a new instance of the data that has the change made. For instance, instead of our imperative example above, where we add 5 to some variable: "$x=x+5$," a functional design would make the update like this: "updated_$x=x+5$."

Additionally, in this design paradigm, functions have no side effects—in other words, **they do not modify any data passed in, nor any other data in the program**. You can think of these as being similar to mathematical functions, like this linear one: "$f(x)=5*x+7$." In this case, $f(x)$ is equivalent to a return value, and x is the input parameter. When you were running through assignments or exams, completing multi-part problems that relied on functions like these, you'd likely think of the function's results solely in terms of generating a new output, which you could then use in the next part of the problem. You wouldn't expect the function to have an effect like "$x+=1$," so that you would have to go back and update the value of x in earlier locations in the problem. The thinking behind functional programming is very similar, and it aims to make code safer and easier to understand by constructing functions that don't touch data outside of themselves. **This has a notable impact on code comprehension overall, as it's easier to understand what's going on in a program just by looking at function names and outputs, rather than having to pore through every function to examine the side effects.**

Functional programming really shines in places with lots of data processing. Let's return to that example of selling puzzles online, and say that we want to add a service for helping customers figure out puzzles they're stuck on. We'll create some data to represent a puzzle, which is a 2D grid of numbers representing pieces and moves. For the purpose of keeping this example high-level, we're leaving some details vague (like what kinds of puzzles these are, and how the solving functions work).

The key difference between this new puzzle data and our imperative Customer class is that we will not modify any state; instead, we will create and return new data from our methods. And when we solve a puzzle, we'll do so by creating a new puzzle array with the correct configuration instead of modifying the previous one. Below, we show a functional implementation to make a move on a puzzle object. We assume some solving-related methods have already been implemented, and reference them below.

Pseudocode for Functional Puzzle Solving:

```
int[][] startingPuzzleState = {{3, 1} {4, 1} {6, 0} … }
// Given a list of moves by the player, return a new puzzle
object with the updated state
        int[][] tryMoveSet(int[][] moves, int[][] puzzle) {
        for move in moves:
int[][] newPuzzleState = puzzle + m; /* creating a new state
by combining
*/ the old and new ones
                if isSolved(newPuzzleState) {
                return newPuzzleState;
                }
// no solutions found, return original state
        return puzzle;
}

// An example puzzle variable

// The player tries some possibilities:
newPuzzleState = tryMoveSet(3, 5, startingPuzzleState)
```

We can see that the function tryMoveSet doesn't modify any data passed into it; it only creates new instances from that data. By sticking to a functional design, a developer can test various solving methods on puzzle objects without worrying about how their state changes as a result.

This design also tremendously helps concurrency. If you were to find yourself wanting to test a number of puzzles and various solving methods at once, you could parallelize running these functions on the starting puzzle states without worry of conflicts, as the functions will never modify the given data. Finally, updating your solving code becomes much simpler, as you know the precise effects of changing the implementation of a single function. You can even exchange some functions for new ones entirely, as doing so will have no unexpected state effects on the program.

This entry only brushes the very surface of functional programming—there are many other notable traits of common functional languages, including composition, tail recursion, and currying. Many languages that feature these traits, like Scala and Clojure, are hybrid and allow mutable data structures, allowing programmers to make decisions about which paradigm is most helpful under varying circumstances. Notable imperative-style languages like Python and Java also feature functional capabilities like lambdas, unnamed functions you can pass as variables. If you're interested in learning more, we highly recommend *Functional and Concurrent Programming: Core Concepts and Features* by Michel Charpentier, which is the textbook used in one of the authors' programming language classes and found to be very thorough.

FURTHER READING

https://stackoverflow.com/questions/1619834/what-is-the-difference-between-imperati
    ve-and-procedural-programming-paradigms

WHY CAN'T 0.1 BE REPRESENTED ACCURATELY IN COMPUTING SYSTEMS?

Have you ever tried testing whether two floating point numbers are equal in a program, only for it to turn out false when you knew they should have been the same? How about trying something explicitly—(0.5 − 0.4 == 0.1)? This should be true, right? But if you try it, you will get false, and if you print it out in, say, a Python terminal, you'll get...

```
>>>> print(0.4 - 0.3)
0.10000000000000003
```

You'll see similar behavior without even having to do any arithmetic. Continuing in Python, if you try something like this:

```
>>> test_number = 0.1
>>> f'{test_number:.20f}'
```

You'll get the following output:

```
>>> '0.10000000000000000555'
```

It seems like such a simple, small value to represent, so why do we get these oddities with extra numbers at the end? You can even try manually setting **test_number** equal to "0.1" with 20 trailing zeros, and printing the value will still show the same output. Go ahead, try it!

This is because 0.1, or 1/10ths, isn't actually representable in the base two number system, or binary, which forms the low-level storage of all modern computing systems. For more info on binary, check out our other entry on the base

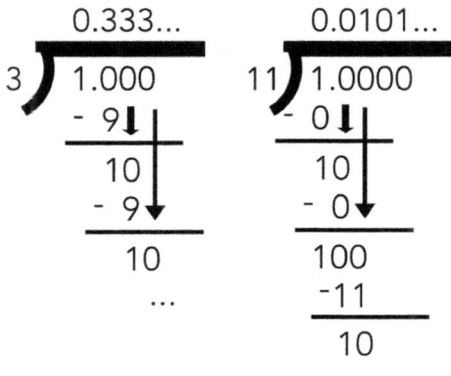

Long division of 1/3 in decimal and binary.

2 numbering system. The same problem actually happens with some decimals in base 10! Let's look at an example.

Consider the fraction ⅓. When converting this to decimal format, the result is 0.33333.... Or 0.3 repeating. If you try to divide out the fraction on paper, you'll quickly see that the division goes on forever, as you always end up attempting to divide 3 into 10. In binary, the same happens when attempting to represent 0.1, whose equivalent fraction is 1/10. In binary, this fraction is 1/1010, which results in a repeating decimal. This means that at the bit level, in program memory, the number must be cut off and rounded to be stored, and so we end up with extra small digits at the end of our floating point values.

The web article https://www.exploringbinary.com/why-0-point-1-does-not-exist-in-floating-point/#google_vignette includes a terrific explanation of this as well, and is worth checking out for more explanation on why this happens.

WHEN PROGRAMMING, WHY ARE FLOATING POINT OPERATIONS SO TRICKY?

There are a couple of aspects to this: one being how floating point values are represented at the bit level, behind the scenes; and the other has to do with how oversized values, and variations in precision, have to be handled as a result. For more details on the underlying representation, check out "Why can't 0.1 be represented accurately in computing systems?". The short version of that is two things: several numbers which can be represented in a fixed-size decimal, like "0.1," can't be in binary (which is 0.000110011... repeating), and floating point representation in computer systems has a fixed size. Repeating decimals must be rounded, and this rounding introduces some inaccuracies that can cascade across floating point operations. Furthermore, there are two floating point types available across most popular programming languages—floats and doubles—which have different sizes, and thus levels of precision. This also impacts the effects of rounding and accuracy across floating point arithmetic operations. Let's break all of this down into some examples:

As we mentioned in that other entry, a number that presents a repeating decimal in base 2, say, 0.3, will be rounded to fit into a fixed set of bits. In addition to affecting the expected value of a number (setting a variable equal to 0.1 will actually set it to "0.1...00555"), this also means that basic operations like addition, subtraction, and multiplication on floating point numbers are no longer associative. In other words, they can no longer be freely rearranged without affecting the final value. For instance, look at this code output in Python when we try adding the same three numbers together, but in different orders:

Code	Output
```>>> addition_order_1 = 0.1 + 0.2 + 0.4```	`>`
```>>> addition_order_2 = 0.1 + 0.4 + 0.2```	`>`
```>>> print(f'{addition_order_1:.20f}')```	`> 0.70000000000000006661`
```>>> print(f'{addition_order_2:.20f}')```	`> 0.69999999999999995559`
```>>> addition_order1 == addition_order_2```	`> False`

Even though we're adding the same values, the final output is slightly different. In the first ordering, the result of 0.1 and 0.2 is rounded before being added to 0.4; in the second, the result of 0.1 and 0.4 is rounded. Although the values are pretty close and they're comprised of the exact same numbers, they're still not fully equivalent, as we can see in the equality test output. This behavior trips up a lot of programmers who try to perform comparison tests on floating point numbers! As a workaround to direct equality tests, programmers will often use a "tolerance value," or some very small number (for example, 1.11e-16, and test whether the floats are close enough that their difference is smaller than that value).

Another difficulty has to do with the fact that many languages have different precision options, generally termed "doubles" and "floats." Floats are typically 32 bits in size, and doubles are 64 bits. In Python, there is only one size for floating point values; they are all 64 bit. But languages like Java and C++ support these different formats, which unsurprisingly have an effect on the stored value as well. Let's look at another example, using two different precision levels:

Code

```
#include <stdio.h>
int main()
{
    double test = 0.1;
    float float_rep = 0.3;
    double double_rep = 0.3;
    printf("\nfloat value for 0.3 is: %0.20f",
float_rep);
    printf("\ndouble value for 0.3 is: %0.20f",
double_rep);
    printf("\nSum for float plus double is is: %0.20f",
(test + float_rep));
    printf("\nSum for double plus double is: %0.20f",
(test + double_rep));
        return 0;
}
```

Output

```
float value for 0.3 is: 0.30000001192092895508
double value for 0.3 is: 0.29999999999999998890
Sum for float plus double is: 0.40000001192092893287
Sum for double plus double is: 0.40000000000000002220
```

Once again, a direct test for equality here will fail. Also, there's the question of what size constants are: is "0.67" in a line such as "float x=0; x += 0.67" a double or a float? Care must be taken to look up the answer and determine where errors may arise.

To add even more confusion to the format, floating point values also have two possible representations for 0, infinity (when the number is simply too large to be represented in digital format), and invalid numbers (those caused by, for instance, trying to divide by zero). This is because, similar to integers, a single bit is used to represent whether a number is positive or negative. You can bet that these cases also fail equality tests—+0.0 is not the same as −0.0, even though on paper math, they are equivalent.

These variations can cascade into considerable effects. A popular software developer, Julia Evans, crowdsourced a number of great examples related to floating point errors on her blog, including those easy to encounter in popular video games like Minecraft and Kerbal Space Program. We link to it here for more reading, but we'll summarize a couple of examples: in Minecraft, there used to be a region called the Far Lands, beyond an extremely long distance from the center of the map. Minecraft's map self-generates as one travels outward, and uses something called noise generators to build the terrain with varying structures and biomes. Past a certain distance, the floating point values used for generation would overflow, and the map would begin developing huge swaths of floating blocks of terrain. Going out even farther even messed with the physics of the game and how it calculated block solidity and transparency—some objects and terrain types like stone and dirt were no longer solid, allowing the player to fall through, and regular items no longer functioned properly. This glitch was patched in recent versions of Minecraft, but it's still easy to find images of it through online searches or to explore it by using an older version of the game.

Floating point values are pretty integral to everyday calculations, but they're difficult to represent on a digital system, and have subtleties to their implementation that catch programmers off guard all the time. Hopefully, this entry has given an idea of why some of these difficulties exist, and that the reader will be better prepared to work with them after reading this!

FURTHER READINGS

Goldberg, D. (1991). *What Every Computer Scientist Should Know About Floating-Point Arithmetic–* (March, 1991 Computing Surveys.). Association for Computing Machinery, Inc.

Julia Evans Blog: https://jvns.ca/blog/2023/01/13/examples-of-floating-point-problems/

HOW DO GARBAGE COLLECTORS IN PROGRAMS KNOW WHICH OBJECTS TO DELETE?

Garbage collection (GC) tends to fall under two general areas: mark and sweep, and reference counting. Both have their benefits and drawbacks.

Remember that dynamically allocated objects are stored on the **heap**, while local variables—for instance, variables you declare with just "int X;" or "char this_letter;"—are stored on the **stack**. Anything declared as a global variable is stored in a dedicated area of program memory, which lasts throughout execution. Garbage collectors only manage dynamically allocated memory—therefore, they only clean up things out of the heap.

Mark and Sweep

With mark and sweep, garbage collectors will "sweep" the entire active program memory, including the stack and global memory, for any values that look like they're on the heap. There's a tremendous amount of theory that goes into determining "looks like they're on the heap," but a simple example is that the value may be a number that falls within the heap address, indicating it's a pointer to some allocated memory. If detected, the corresponding address on the heap is marked "reachable." Depending on the garbage collection implementation, the GC might also completely freeze execution so that it can sweep registers as well. At the end of a mark and sweep cycle, any objects left unmarked are deleted and their corresponding area of heap memory is marked free for future use.

Let's demonstrate this with a very small pseudocode-based example, listed below. We have a program that calls a function to dynamically create an object and set an integer value. It doesn't return that pointer, though, so the memory is left inaccessible—hence the name "badAllocateMemory." In languages like C or C++, this would create a memory leak, but here, we have a garbage collector. Below, we show the stack variables for these functions (we leave out frame information from the stack, just for easier reading). "SP" refers to the stack pointer. If you're not familiar with what this is: in very short terms, it's a marker that keeps track of data that has been pushed on and off the stack. If you have no prior exposure to assembly or what the stack is, it's worth taking a moment to search for a brief overview of the stack and stack frames online.

```
1 Main() {
2   badAllocateMemory(); // create something on the heap
3   return;
4 }
5 Function allocateMemory() {
6   Object current = new Object(); // create something on
the heap
```

```
7    int test = 3; // create a stack variable
8    return; // The object memory is now inaccessible!
9    }
```

We begin our example at line 6; we've started the program, called the function to badAllocateMemory, and are about to allocate a variable. The stack for this function scope has no variables yet. Bolded addresses on the stack show the "active," accessible areas on the stack.

State of Memory at Line 6:

Stack Addresses	Values
0x00001F8	// data associated with main

Heap space	

After executing lines 7 and 8, we have a pointer to an object on the heap, and a local variable:

State of Memory at Line 8:

Stack Addresses	Values
0x00001F8	// data associated with main
Start of badAllocateMemory()	
0x00001e8	**0x00077ffe8; current**
0x00001A8	**0x00000003; test**

Heap Addresses	Value	Is Accessible
0x00077ffe8	**Instance of Object**	**Yes (referred to by stack address 0x00001e8)**

If the garbage collector were to kick in at this stage, it would see that the "current" allocated memory is still within active local memory and mark it in the heap as being accessible.

After executing the return statement, we are back in main at line 3. The stack pointer has been adjusted to reflect that our local context only includes the variables in main. Although the values from allocateMemory() still exist on the stack, they're inaccessible; when the garbage collector sweeps the stack for local references, it will only sweep as far back as the current context and above(main). A real-life comparison is tearing some old notes out of your binder and throwing them in the recycling: the writing still exists, but is no longer accessible in your "active" working area, the binder.

Line 3:

Stack Addresses	Values
0x00001F8	*Data associated with the main function...*
Bounds of main function stack; addresses below are no longer accessible	
0x00001e8	*0x00077ffe8; current*
0x00001A8	*0x00000003; test*

Heap Addresses	Value	Is Accessible
0x00077ffe8	Instance of object	No (no active references found)

While effective, the downside of this is that it's time-consuming to sweep all program memory for references. Variations of more efficient mark and sweep include things like waiting until memory fills up to perform a round, and only checking swept values against selected objects on the heap based on age and usage.

Automatic Reference Counting

The second method we mention here, automatic reference counting, approaches things from the opposite direction: every piece of memory at allocation is given a tally. Whenever a reference to that memory is initialized, the tally is increased. When a reference is removed, the tally is decreased.

Although this is faster than mark and sweep, it has some pitfalls. For example, if two objects A and B contain references to each other are cleared at the same time, their underlying heap memory might never be freed during the life of the program. When trying to free memory, the GC will see that A points to B, and thus B has a tally of 1, and will keep B in memory. But because B points to A, A's tally is also at 1—and so A is kept in memory as well. This is called a circular reference, and is a type of memory leak! Some reference-counting–based garbage collectors like Java's and Python's periodically run an algorithm similar to mark and sweep to detect and remove these, as otherwise, this memory remains stuck for the duration of program execution.

GC is a tremendous aid to programmers, saving time and errors in handling inaccessible memory. Years of research and varying implementations have gone into making it as efficient, smooth, and seamless a process as possible for developers. Under the hood, though, the fundamental concepts are simple: if no references to a given object can be accessed anywhere in the program, then it needs to be deallocated and freed.

For more info, we recommend *The Garbage Collection Handbook: The Art of Automatic Memory Management* by Richard Jones.

FOR ARITHMETIC EXPRESSIONS, WHY DO SOME PROGRAMMING LANGUAGES NOT FOLLOW OPERATOR PRECEDENCE UNLESS YOU EXPLICITLY ADD PARENTHESES TO OPERATIONS? WHY DOES ORDERING VARY SOMEWHAT ACROSS DIFFERENT LANGUAGES?

Before your program's mathematical expressions can be executed, they must be lifted from source code into your compiler or interpreter in a step called parsing. Typically, this involves using a set of syntax rules to scan through source code and recognize expressions and statements. Parsing accurately and flexibly is a somewhat challenging task, and in the case of mathematical expressions, the order in which code is parsed into statements also affects the order in which they're executed. Although humans are good at chunking text-based statements into components that make logical sense (like reading a mathematical expression and determining what to calculate first), compilers can only scan things one character at a time, forming sub-expressions as they recognize them, and then performing those calculations in order. To handle any ambiguity, parse rules are usually defined with a strict ordering that determines their priority. For various design-related reasons, different languages often define their parse rules and ordering in varying ways (and some languages don't follow any kind of mathematical order of operations at all!), which gives rise to some confusion when programmers switch languages.

Compilers, as the last stage of their process, must generate machine-level instructions that correspond to user code. For instance, if you were to write some code to convert Fahrenheit to Celsius, like this, where the variable **fahrenheit** has already been declared and initialized elsewhere:

```
int celsius = (fahrenheit - 32)*5/9;
```

Then, correspondingly, your compiler must generate a sequence of binary instructions, something like this:

1. subtract 32 from memory location for **fahrenheit**
2. divide 5 by 9
3. multiply the result of step 2 by memory location for **fahrenheit**

If you're using an interpreted language instead, like Python or Lua, then something similar happens: the interpreter will take that basic code and execute these arithmetic operations dynamically, instead of creating a binary that contains those instructions. But there's a bit of a gap between the user code and the corresponding backend actions, which is where parsing comes in. Before your compiler can generate the binary code for these instructions, it must be able to recognize them in your source code. In general, the order in which statements are recognized (or parsed) also affects the order in which they're executed.

Let's take a look at an example of parsing from a somewhat high-level, but still practical perspective. Let's say we're trying to write an interpreter for a simple programming language with some basic math operators. In our interpreter, we've already written the backend code, which can take an expression and calculate the result for it, but we need to write a parser to scan some input code and extract expressions for the interpreter from it. We'll write some parser rules to do so. For this exercise, we'll assume that our parser scans input text from left to right, and we'll use a format called "Backus-Naur Form," or BNF for short. BNF is a very common format in which to write parse rules, and it consists of a set of rules and their definitions, also called derivations. These are typically written in the format of a rule name, and its definition, which describes the format and structure of the rule. When a parser scans through source code, it will test if the input matches any rule definitions, extract that info, and give it the corresponding name.

So, how might one define the format for a mathematical expression in source code? We can start with a simple operation: addition. This isn't too difficult: addition often takes the form of two numbers and a plus sign, like this "5+9." Let's assume we already have a rule to recognize numbers, which is filed under the name "NUM." So, we can add a rule to our grammar correspondingly:

ADD::=NUM "+" NUM

Note that typically, there is also a stage of parsing called lexing, which is designed to recognize tokens. Tokens are things like numbers, strings, and punctuation marks. For the sake of simplicity in this entry, we're assuming we already have a lexer, and correspondingly, a way to detect numbers and operators. With the above rule added, when the parser scans some source code "5+9," it'll recognize that this is an ADD operation consisting of two numbers, 5 and 9, and will pass this info to the interpreter to add the two numbers together. There's a couple of problems here, though: if we try adding a longer set of values, the parser won't recognize the last one as being part of a valid ADD rule:

"5+9+3"

"5+9" -> ADD

ADD "+3" -> error!

And even if we tweak this to count as technically valid syntax:

"5+9+30+1"

We'll end up with two separate ADD operations: one for "5+9," and one for "30+1," which will be passed to the next component of the compiler as "ADD (5, 9) and ADD(30, 1)." We can add support for multiple options, called candidates, using the vertical bar operator:

ADD::=NUM "+" NUM | NUM "+" NUM "+" NUM | NUM "+" NUM "+" NUM "+" NUM |...

But this will be incredibly cumbersome, and only works up to a limited number of operands. Fortunately, we can tweak our grammar a bit to add recursion, where a rule refers to itself, so that we can stack repeat operations against each other.

ADD::=ADD "+" ADD | NUM "+" ADD | NUM "+" NUM

When the parser now hits an expression like "5+3+2," the expression will match the second candidate NUM "+" ADD, and then the parser will recursively explore the candidates for ADD once more to detect that "3+2" matches the candidate "NUM+NUM." How does it know when to stop? Parsers will explore candidates until there are no more "non-terminals" left, or in other words, until there are no more rule tokens that can be expanded into more rules. Oftentimes, textbooks and other resources describing parsing rules will use a format called an Abstract Syntax Tree to better visualize them. We'll employ them here as well to visualize some of our rules as we go along.

We'll go ahead and add one for multiplication, too, called "MUL":

MUL::=MUL "*" MUL | NUM "*" MUL | NUM "*" NUM

But we still have a major, glaring issue: we don't have the ability to mix operations! Our parser won't be able to handle expressions like "5+3 * 8." Let's reformat our rules a bit to better take advantage of the recursive and flexible nature of BNF, and define a new kind of rule that encompasses both operations and numbers alike:

EXPR::=MUL | ADD | NUM

ADD::=EXPR "+" EXPR

MUL::=EXPR "*" EXPR

The rule "EXPR" can evaluate to an operation or a number. This works for our purposes because although these seem like very different things, they are both valid operands to an operation. Think of the expression "5+10 * 3": this can be expressed as "ADD (EXPR, 5)," where EXPR is equivalent to MUL (10, 3). Now, with these very short rules, we can evaluate statements as simple as "5+9" or even more complex than the one we just described. There's no need to detail all the possible combinations of values that comprise an expression—instead, the parser will recursively unpack the various components until there are no rule candidates left. If we want to add other operations, it's similarly very simple to do so.

Note that the order in which we define our rule candidates is important. In general, parsers try the first candidate available, and then the next, until it find one that fits. In many cases, multiple candidates could match the same rule, so it's important to consider how one candidate choice affects the ones that come after it. For instance, if we defined EXPR as "NUM | ADD | MUL" instead, placing the number first, then our parse result for "5 * 10+6" would end up like this: NUM, and then separately "* 10+6," which wouldn't match any rules.

At this point, we've determined we can parse expressions correctly, but what values do we get when we execute them? Let's try "5+9 * 2":

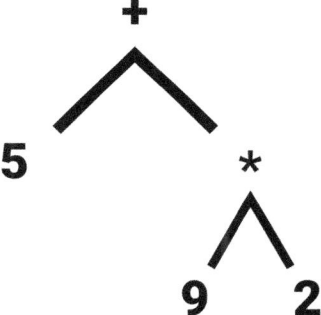

If we were following order of operations, we'd expect to have 23 as a result, but instead we get 28! Recall that parsers explore rules in order, and we put our addition and multiplication tokens at the same level, so they have the same priority. They get evaluated in whatever order they appear in the source code. If we want to make sure multiplication and division get a higher priority, and always get evaluated first, we need to make sure our higher-priority operators appear first:

EXPR::=MUL | ADD | NUM

ADD::=EXPR "+" EXPR

MUL::=EXPR "*" EXPR

Now, when our parser scans the expression "5+9 * 2," it will initially parse it as EXPR * 2, and this will be passed to the interpreter and computed first. Most popular languages encode operator precedence like this, and so addition and subtraction have a lower precedence than division and multiplication.

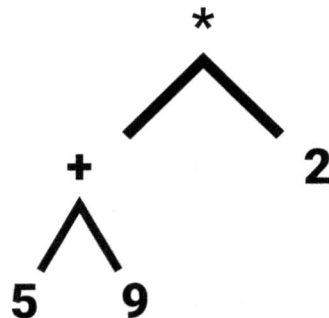

To allow programmers to avoid any ambiguity in executing expressions, language parsers often define their parentheses-matching rules near the top of their priority list. This way, programmers can have some certainty in explicitly declaring how their expressions will be evaluated, without having to understand the underlying parse rules or compiler processes underneath them. As any good programming instructor or textbook will tell you, the use of parentheses is always a good idea in your code to ensure that there will be no surprises in your code's output.

Hopefully, this entry has shone a light on why and how languages vary in their order of operations (and also why it's of tremendous importance to use parentheses any time the order of operations may be ambiguous, when you're not sure what the underlying precedence is in the language you're using, or if your code might be read by someone else who is unfamiliar). We've only lightly touched on the concept of parsers, language grammars, and evaluation order, and there are numerous types of parsers (top-down, bottom-up, recursive descent, and so forth). If this topic interests you, we highly recommend the following resources:

If you're pondering how compilers assess things like unresolved references, the usage of undefined variables, type information, and other complexities: this kind of analysis works by running multiple compilation passes. For instance, if the parser encounters a reference to a user-defined object type before its definition, it will leave this portion of the bytecode temporarily undefined. Once all text has been parsed and the first round of compilation is done, the compiler will perform a second pass to complete all undefined locations, now that it has all definitions and other information.

To learn more about parsing and compiler theory, we recommend: *Crafting Interpreters* by Robert Nystrom and *Compilers: Principles, Techniques, and Tools* by Alfred Aho and Ravi Sethi.

ANTLR is a free, Java and C++-compatible library for writing your own parsers and lexers, and it has an interactive webpage where you can immediately try playing with your own grammar rules at http://lab.antlr.org/.

ARE RANDOM NUMBER GENERATORS TRULY "RANDOM"?

Random number generators take many forms, not all of them computational. Shuffling a deck of cards or flipping a coin are forms of random number generation that have been used for millennia. The focus on random number generation is not just in the system, but also in the environmental factors that go into it: a coin flip can be skewed by the force of the thumb underneath it, the weight of its sides, the surface it falls on, and so forth. After a certain number of hands, you may recognize if a deck of cards has not been shuffled well. The person or thing that "randomized" it is a big determinant for its efficacy.

While something may not be randomized "well," the randomness of the result is also quantified by how easy it is to replicate the exact result per generation. With randomness generation, environmental factors should be such that the number generation cannot realistically be quantified and reproduced.

In truth, "random" number generators—ones found in software algorithms—are usually only "random" during a certain period. The same random number generator used to create a password would not be the same one for a cryptographic algorithm. Instead, these generators are often called "pseudorandom," and they have varying degrees of randomness per use case. This is because a pseudorandom number generator will, after some volume of sequences, begin to repeat itself.

A pseudorandom sequence generator is designed around an algorithm that operates on some initial value. The point at which it repeats is referred to as its period, and the value which the generator operates on is called its seed. Number generation is necessarily deterministic, as with everything a computer does, and generation takes some degree of computational power respectively. Therefore, creating numbers for something like a virtual dice roll takes a relatively small frame of power. You would ideally like to run a dice roll on any computer you run the program on, without having to delve into the world of supercomputers or remote access to them. On the other hand, for neural networks or cryptographic testing, you would need a much larger degree of randomness. In fact, you may need to harvest certain environmental factors that contribute to the effective realization of randomness, possibility, or unpredictability. This is called entropy, which is too complex to delve into for now.

The repetition created from an algorithm may not be immediately identifiable—the pattern, or its outward predictability, could be obscured by some varying degree of complexity in the algorithm itself. However, if you repeat a generator enough times, then certain strings and patterns will emerge over time. This is partially why some cryptographic algorithms, for example, hashing algorithms, do not rely solely on random number generation. The point of pseudorandom number generation is to create the veneer of randomness, so that for all practical purposes, the output appears random. It does not matter whether the randomness is actually "true," as long as the output cannot predictably be replicated by repeat tests or input values.

FURTHER READINGS

https://blog.cloudflare.com/why-randomness-matters/
https://engineering.mit.edu/engage/ask-an-engineer/can-a-computer-generate-a-truly-random-number/
Viega, J. (2003). Practical random number generation in software. In *19th Annual Computer Security Applications Conference, 2003. Proceedings*. IEEE.

WHY ARE TIME ZONES DIFFICULT TO PROGRAM?

At a global level, times across time zones can be difficult to reconcile programmatically. There are a large number of edge cases for time zones and, what's worse, many of those change over time. Additionally, wide variances have occurred over human history as to how time and time zones were and are managed, so historical times are even harder to manage programmatically.

Many of the difficulties with time zones arise, in essence, from the passage of sunlight and nightfall crossing the surface of the rotating, imperfect spheroid earth in its journey around the sun. While humans try to maintain 24 master time zones in a 24-hour day—where some of these are even, in some areas, subdivided into half-hour offsets (for example, India)—across 365 days for a year, it still requires the occasional adjustment. Additionally, the rotation of the earth does not match uniformly back to the 24-hour clock we use to measure the passage of time. (Also, Earth actually has a year that's an average of 365.24 days per year.)

Then governments, people, and laws get involved and time zone conversions get to be much more complicated. For example, while there are 11 time zones in Russia, its southern neighbor China has 1 Chinese time zone (resulting in some very odd edge cases there). Both the United Kingdom and the United States use daylight savings time but they do so starting and ending on different dates (UK daylight time starts before US daylight time, creating an additional offset between all of the time zones in the two countries).

But even the time zone within *a single place* can be difficult to program.

Let's localize to what looks to be a reasonably simple exercise: time zone conversion within the state of Arizona. We will map to the four time zones of the continental United States. The United States has a total of seven time zones: four continental ones and three additional which apply only to Hawaii and Alaska. Time zones largely alternate between "Standard" and "Daylight" modes, seasonal adjustments made purely by fiat of law—there is no intrinsic change in the nature of time or space, just in the laws in effect concerning the time keeping.

The state of Arizona uses Mountain Standard Time (MST) year-round, or at least it has and, as of this writing, has continued to do so, since 1968. That is, residents of Arizona do not adjust their clocks for the seasonal changes as does the rest of the nation (so Daylight Savings Time is not observed or used in the state). Arizona thus falls out of sync even with other states along the Rocky Mountains that are using Mountain Time (when those other states switch for 6 months out of the year to what is called "Mountain Daylight Time" or MDT).

However...

Arizona's avoidance of MDT *does not include* the land in Arizona of the Sovereign Nation of the Navajo people. The Navajo Nation, which occupies some 27,000 square miles and was established in 1868, *does* observe MDT. A further complication occurs within the Hopi Reservation in Arizona—some 2,500 square miles and entirely *contained within* the Navajo Nation—as the Hopi Reservation there *does not* observe MDT.

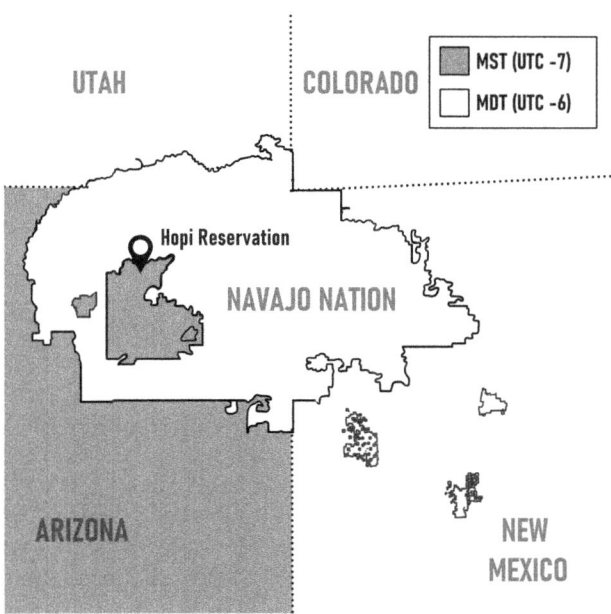

There are edge cases like these everywhere, and every one of them must be tracked and captured correctly in code to prevent discrepancies and conversion issues. Some regions still change their time zone—Samoa changed its official time as recently as 2011! Given all the variations, changes, and edge cases, tracking time zones is no simple task.

WHY ARE LANGUAGES THAT DON'T HAVE AUTOMATIC MEMORY MANAGEMENT LIKE C STILL SO POPULAR?

C is still used extensively in areas that have very heavy constraints around performance, memory usage, or the ability to control memory and hardware addressing. In the real world, these areas can range from monitoring software to embedded systems and backend operating systems code. C's manual and rudimentary design are actually a strength in all these areas!

Features like garbage collection (GC) while adding a tremendous amount of safety to programming languages, also tend to add delays to program execution. This isn't perceptible to the average developer writing user-facing applications, but on the operating system level and embedded systems, the slowdown can be apparent. Things like scientific monitoring devices also often need to be extremely precise, and the addition of a few milliseconds at every memory operation can seriously add up and impact the quality of their results.

In most cases, the benefits of these memory management features outweigh the delays. Programmers using them are far less likely to make mistakes involving null references and memory overwrites. In newer languages like Java and Python, memory management is so well-embedded into the language that pointers are not even present as a feature of the language at all. For their general purposes, there is no need to make the mechanics of pointers available to the user. They're still present on the backend (ultimately, programs must allocate and deallocate memory *somewhere*), but the compiler or interpreter does a tremendous amount of work to handle them for the user automatically and invisibly.

For example, when working with arrays of variable sizes in C and C++, users must manually allocate memory. If the user then needs to increase the size of a dynamic array, they must create a new array with the updated size, copy everything over, and destroy the memory for the old array. In languages such as Python or Java, when the user needs to expand the size of a list, this same procedure still happens, but the interpreter takes care of it all in the background. All the user has to do is call the correct function to append to their list, and the interpreter does the rest. Depending on the type of array used, languages like Java and Python will even create an array slightly larger than what the user requested in case they later want to add items, because creating a new one from scratch and copying everything over is quite time-intensive. But of course, this means that the array takes up slightly more space than expected, without any of the user's control, and is slower to work with—all the checks that the language does on the backend to ensure memory safety take extra time. In high-performance systems, it's more helpful to users to be able to perform the exact operations they need, when they need to, and be aware of exactly how much memory they're using at all times.

Thus, C is still in widespread use today despite the existence of many languages that are much easier and safer to write in. Not only that, but many languages have libraries to run C code for modules that need better speed and precise memory management! The Python interpreter itself is written in C, and the Python Numpy library, a comprehensive math library, is partially written in C and C++ for certain computation-intensive operations. C is not going away any time soon!

WHY DON'T MODIFICATIONS TO FUNCTION PARAMETERS PERSIST ONCE THE FUNCTION RETURNS? WHY MUST THEY BE PASSED IN AS POINTERS TO BE MODIFIED PERMANENTLY?

This question is referring to the case where you're passing some primitive (integer, character, integer array, etc.) variables into a function and then modifying them. If you're very far in your programming classes, or already working in software development in some capacity, you're probably aware that any changes you make to those variables won't stick—as soon as the function returns, they revert to their original values. The exception here is if you're working with a language like C or C++, which allows you to work with pointers; in that case, you can instead do something called pass by reference, and work with the variable as a pointer to get the change to persist. But why is this the case? The short of it is that it has to do with the limitations of program memory when entering and exiting functions. At the hardware level, info about your program's local variables, the current scope or function space, and where to return from a given function is stored on the stack. When you enter a function and pass in variables by value, what happens in program memory is that those values are copied into a local space designated for that function. Although it looks like you're modifying the same variables that exist in higher levels of the program, you're actually modifying copies of them in the local space, and these copies are lost when the function exits.[1]

This answer assumes you have some understanding of stack versus heap memory and an awareness of what registers are. Recall that local variables are stored on the stack. When a variable is passed into a function, it is passed in either by registers or by pushing a copy of its value onto the end of the stack for the function to retrieve when it begins. When the function returns, it restores any data in registers that it overwrote, and resets the active area of the stack to correspond to that of the function that called it. This procedure sounds rudimentary, but it's extremely important for making sure that there are no unexpected side effects during program execution. Let's take a look at some examples, using the code sample below, which simulates adding items to a shopping cart and calculating the user's current running balance. In this code, we have a function to calculate and return a new balance based on the cost and discount passed in. For ease of illustration, we use a convention where variables are passed in via the stack.

Shopping Cart Program

```
int getNewTotal(int currentTotal, int discount, int cost)
{
    int actualCost = cost - discount;
    return currentTotal + actualCost;
}
// some other code in main...
// previously initialized: userBalance = 45
userBalance = getNewTotal(userBalance, 10, 30);
printf("\nYour total is: %i", userBalance); // prints 65
```

What's happening on the stack at the same time?

We start off in main, with the userBalance variable on the stack already:

Stack Address	Stack Values
0x0002	**45 // userBalance**
0x0003	
0x0004	
0x0005	
0x0006	

When we enter getNewTotal(), a new space for it is designated on the stack, as well as its local variable actualCost, and some info about where to return to after it exits. We can also see the variables placed on the stack as well.

Stack Address	Stack Values
0x0002	45 [variable: userBalance]
	Start of stack region for getNewTotal()
0x0003	**[Address in main to return to]**
0x0004	**45 // the value of userBalance**
0x0005	**10 // parameter: discount**
0x0006	**30 // parameter: cost**
0x0007	**20 // new local variable: actualCost**

The bolded text represents the active area of the stack.

When we return from the function, our active stack area goes back to just including data from main, and the line userBalance=getNewTotal(userBalance, 10, 30) stores the new total returned from the function into stack address 0x0002, which corresponds to the userBalance variable.

Stack Address	Stack Values
0x0002	**65 [userBalance]**
	Old, inactive stack region (will be overwritten by the next function call)
0x0003	*[Address in main to return to]*
0x0004	*45 // the value of userBalance*
0x0005	*10 // parameter: discount*
0x0006	*30 // parameter: cost*
0x0007	*20 // new local variable: actualCost*

After this, the stack addresses previously used by calculateNewTotal() will be used by the call to printf shortly after, and completely overwritten.

Notice that at this level of representation, there's no way to indicate what value corresponds to which variable name. The new value ends up being stored in the correct place because the C or C++ compiler keeps track of this variable info and generates all the assembly instructions to do so for us. We know intuitively, looking at the stack here, that the value stored at address 0x0004 is from the userBalance variable, but from the machine's perspective, it only knows that it is operating with some function that has three parameters and one local variable, and the addresses for each of them.

Now, let's get to what happens when we make the mistake of trying to modify a passed-in variable! We've now got a slightly modified version of our program, where we try to skip returning the updated total, and instead adjust the parameter directly:

```
               Bugged Shopping Cart Program
int getNewTotal(int currentTotal, int discount, int cost)
{
    int actualCost = cost - discount;
    currentTotal += actualCost;
}
// previously initialized: userBalance = 45
userBalance = getNewTotal(userBalance, 10, 30);
printf("\nYour total is: %i", userBalance); //
Error - this prints 45 instead of 65!
```

Stack Address	Stack Values
0x0002	45 [variable: userBalance]
	Start of stack region for getNewTotal()
0x0003	[return address]
0x0004	45 // parameter: currentTotal
0x0005	10 // parameter: discount
0x0006	30 // parameter: cost

And here's what happens on the stack when we run this, enter getNewTotal, and execute currentTotal += actualCost:

Stack Address	Stack Values
0x0002	45 [variable: userBalance] // this is unchanged by the modification
0x0003	[return address]
0x0004	65 // parameter: currentTotal // <- we end up changing this parameter!
0x0005	10 // parameter: discount
0x0006	30 // parameter: cost
0x0004	45 [parameter: currentTotal]

We're modifying the value stored in our current local scope! After the function returns and its corresponding area of the stack is popped off, we are left with our original values:

Stack Address	Stack Values
0x0002	45 [variable: userBalance]
	Old, inactive stack region (will be overwritten by the next function call)
0x0003	[Address in main to return to]
0x0004	45 // the value of userBalance
0x0005	10 // parameter: discount
0x0006	30 // parameter: cost
0x0007	20 // new local variable: actualCost

You may be thinking, "Is there a way to specify that we want to change the original variable we pass in?" And this precisely what passing in by reference accomplishes. Let's tweak our code correspondingly:

By Reference Shopping Cart Program

```
// change the currentTotal parameter type to be a pointer
int getNewTotal(int* currentTotal, int discount, int cost)
{
    int actualCost = cost - discount;
    *currentTotal += actualCost;
}
// previously initialized: userBalance = 45
userBalance = getNewTotal(&userBalance, 10, 30); // pass
in userBalance by reference
printf("\nYour total is: %i", userBalance); // prints 65,
as expected.
```

And then look at what's happening on the stack:

Stack Address	Stack Values
0x0002	45 [variable: userBalance] // this is unchanged by the modification
0x0003	[return address]
0x0004	0x0002 // parameter: int* currentTotal // address of userBalance
0x0005	10 // parameter: discount
0x0006	30 // parameter: cost
0x0004	45 [parameter: currentTotal]

Now, getNewTotal() expects a pointer, which we retrieve by fetching userBalance's address. You can see on the stack that as a result, when we enter getNewTotal, instead of the value 45, the address 0x0002 is stored on the stack as one of the input parameters. Finally, when we executed *current-Total += actualCost; (note the dereference operator, as now we are working with a pointer), we are effectively saying: "modify the value stored at this other address," and we can see this take place on the stack:

Stack Address	Stack Values
0x0002	65 [variable: userBalance] // Now our original variable is changed!
0x0003	[return address]
0x0004	0x0002 // parameter: int* currentTotal // address of userBalance
0x0005	10 // parameter: discount
0x0006	30 // parameter: cost
0x0004	45 [parameter: currentTotal]

This is why variables seem to revert when passed back in—in reality, they were never changed at all. This is also why you may notice a quirk when working with dynamic variables and objects: if you pass an object into a function, and modify one of its attributes, the change to the attribute will persist after. Dynamic objects are stored on the heap, and are often quite large; the local variable that stores one is actually storing the address of the object in memory. Thus, when you pass a dynamic object into a function, its address is given as a parameter much in the same way that we passed in the address of userBalance in our updated code. Therefore, when you change an attribute, you are actually changing the original object's attribute in heap memory. But if you try to change the object itself by creating a new instance of something and assigning it to the parameter, this change will be lost after the function returns. You're changing the address stored within the parameter, which will be lost on function return.

What about C++ objects that aren't created dynamically? And how does this compare to languages like Java, where pointers don't exist as a language concept, but objects are still stored in dynamic memory?

When you create a C++ object without the use of the new keyword and without any use of a pointer, the entire object is created on the stack, including any attributes. This means that, similar to primitive variable reversion, if you pass the object into a function and then modify an attribute of that object, you're only modifying a copy—the attribute will revert to its original value after the function returns. Passing in a pointer to an object, versus a local object, can have major behavioral differences in C++. In Java, however, every object is referred to by a reference to some allocated memory. There is no concept of dynamic or non-dynamic objects. This means that when you pass an object into a function, and modify an attribute, you are modifying the attribute of the original object in memory, and the change will always persist on function return.

There's quite a bit more detail to these processes that we've left out for the sake of keeping a digestible, high-level summary. But if you're curious and want to learn more, we recommend looking for info about assembly and computer architecture.

FURTHER READINGS

Computer Systems: A Programmer's Perspective, by Randal E. Bryant and David R. O'Hallaron

Reverse Engineering for Beginners, by Dennis Yurichev

https://godbolt.org/—A website that allows interactively compiling and viewing assembly instructions.

https://en.wikipedia.org/wiki/Mountain_Time_Zone

https://www.youtube.com/watch?v=-5wpm-gesOY&start=2

NOTE

1 This actually isn't quite true, but is close enough to the truth for our purposes. For a precise understanding of what happens in program memory, look up info on assembly calling convention and the stack pointer, or see the sources at the end of this section for more info.

Index

Adleman, L. 79
Adobe's multimedia Flash plugin 61
age 84
"Altos-net" 74
American Standard Code for Information
        Interchange (ASCII) table 25–29
    ASCII 127 29
    ASCII A 29
anti-disassembly protections 15–16
Apple 80
ARIN (American Registry of Internet
        Numbers) panel 67
ARPANET 66
assembly basics 12–13
automatic memory management 108–109
automatic reference counting 97

Basic Input/Output System (BIOS) 5, 22
Basic Service Set Identifier (BSSID) 75
Bell's Theorem 69
bitmap graphic 56
Bluetooth 24, 30
booting process 5
bootstrapping 31
breakpoints 9
buffering 18
bug 33–34

C 10, 108–110, 112
C++ 10, 85, 86, 93, 109, 110, 112, 115
capital letters 25
Cathode Ray Tube (CRT) display 6
cell phones and landlines 50–51
central processing unit (CPU) 6
certificate authorities (CAs) 80
Charpentier, M. 89
Clang compiler 85

Clojure 89
cloud storage 42
code sections 13
compilers 85
computer architecture
    64 BIT 2
    anti-disassembly protections 15
    assembly basics 12–13
    base 2 number systems 3–4
    BIOS/UEFI 5
    clock speed 6–7
    debuggers 8–11
    disassembly methods 14–15
    fake conditional jumps 15–16
    junk code 15
    program format 13–14
computer hardware 33
computer memory 22
computing systems 90–91
control codes 27
control-h (backspace) 28
customer class 87

data diskette 41
data storage 42
Dean, J. 71
debuggers 8–11
debugging 13
device identifiers 59
digital graphic formats 56–57
disassembly methods 14–15
distance 83
Domain Name System (DNS)
    IPv4 63
    IPv6 64–65
doubles 92, 93
Dvorak keyboard 40

For Product Safety Concerns and Information please contact our EU
representative GPSR@taylorandfrancis.com
Taylor & Francis Verlag GmbH, Kaufingerstraße 24, 80331 München, Germany

* 9 7 8 1 0 3 2 8 5 6 7 4 2 *